HOW WE SEE THE SKY

HOW WE SEE THE SKY

A Naked-Eye Tour of Day & Night

THOMAS HOCKEY

The University of Chicago Press :: CHICAGO & LONDON

THOMAS HOCKEY is professor of astronomy at the University of Northern Iowa.

The University of Chicago Press, Chicago 60637

The University of Chicago Press, Ltd., London

© 2011 by The University of Chicago

All rights reserved. Published 2011.

Printed in the United States of America

20 19 18 17 16 15 14 13 12 11   1 2 3 4 5

ISBN-13: 978-0-226-34576-5 (cloth)

ISBN-13: 978-0-226-34577-2 (paper)

ISBN-10: 0-226-34576-9 (cloth)

ISBN-10: 0-226-34577-7 (paper)

Library of Congress Cataloging-in-Publication Data

Hockey, Thomas A.

How we see the sky : a naked-eye tour of day and night / Thomas Hockey.

p. cm.

Includes bibliographical references and index.

ISBN-13: 978-0-226-34576-5 (cloth: alk. paper)

ISBN-10: 0-226-34576-9 (cloth: alk. paper)

ISBN-13: 978-0-226-34577-2 (pbk.: alk. paper)

ISBN-10: 0-226-34577-7 (pbk.: alk. paper) 1. Astronomy—Popular works. I. Title.

QB44.3.H635 2011

520—dc22

2011003355

⊗ This paper meets the requirements of ANSI/NISO Z39.48-1992
(Permanence of Paper).

To Yuliana

CONTENTS

PREFACE

The goal of this book is to explain the naked-eye astronomical universe from a point of view on the earth's surface. The emphasis is upon personal experience. (I restrict myself to astronomical phenomena; sky phenomena would include meteorological topics.) A secondary theme is to illustrate some of the ways in which the appearance of the heavens has influenced human civilization.

This is not a "guide to the night sky." Just as a field guide to birds would describe different avian species, and where and when to find them, a field guide to the astronomical sky would describe stellar constellations, and where and when to look for them. On the other hand, a "bird book" would define birds in general, compare and contrast them, and illuminate their behavior. It is this latter format that I try to emulate. Other works (sadly, no longer in print) have attempted to do what I do, but the distinguishing feature of my book is the tie-in with cultural interpretations and practices.

Although virtually all introductory astronomy textbooks cover things that can be observed in the sky without aid, they often do so in abbreviated fashion and with little relation to the historical or societal aspects of these topics. In fact, it is accurate to say that all the subjects in this book may be condensed into twenty pages or less within a traditional astronomy textbook.

My introduction is intended to provide a motivation for what follows. The first four chapters discuss the most plentiful objects in the night sky—the stars. The apparent motions of the stars are the simplest of those seen in the sky. I follow with three chapters on that singularly most important celestial orb: the sun, and its more complicated apparent motion. Chapters 8 and 9 concern the moon, next in cultural importance only to the sun, and whose motions are more complex than that of its daytime counterpart. Chapter 10 explores the visible features on the disks of the sun and moon, while chapter 11 deals with eclipses (events involving both the sun and moon). My last chapter is about planets. Though this is a volume on commonplace astronomical phenomena, I include a few words about ephemeral ones, such as comets and meteors, as well. I conclude with a list of books for further reading.

I thank those who read and critiqued individual draft chapters of this book: J. McKim Malville, Virginia Trimble, Anthony Aveni, Thomas R. Williams, Roslyn M. Frank, Marvin Bolt, Eugene F. Milone, Donald W. Olson, John Steele, Richard Baum, and Nicholas Campion. The following helped me with my anthropology: Ed Barnhart, Bryan Bates, Chris Bennett, Len Berggren, Todd W. Bostwick, David Dearborn, Tom D. Dillehay, Bradley T. Lepper, Bradley E. Schaefer, Rolf M. Sinclair, and Ivan Šprajc. I also am indebted to the useful comments made by two anonymous referees who read the entire manuscript. Ruby Hockey reviewed and improved the writing and grammar from page one to the end.

Unless credited otherwise, all artwork is by Christine M. Tarte.

INTRODUCTION

Why does the world need this book? Why should you want to know about what lies between the pages of this book? These are the questions I will try to answer in this brief introduction.

First: Look around you. Up to half of everything you see is sky. The sky literally surrounds us. Even "The fool on the hill sees the sun going down / And the eyes in his head see the world spinning 'round."[1] Yet few of us know much more about the sky than that.

I think that this is a shame. For the sky is universal. Imagine yourself magically transported to a bar stool in some far-off land. You do not know the local language or customs, much less the cuisine. Nevertheless, there is one thing that you do have in common with the stranger seated beside you. You both have lived your lives under much the same sky.

Yet there is a difference: The more attuned you are to modern, Western society, the less likely you are to know about your sky. As I sat in the *ger* (yurt) of a family living on the Mongolian steppe, supping on kindly offered cheese and yogurt, I routinely asked the same question. I knew that the door of their nomadic house always faced south. My question was, simply, how do you know where south is? The answer was always the same. They just *know*. I was asking a question so obvious, that it was as if someone had asked me how I know one plus two equals three, or how I know how to spell my own name. This knowledge is so quintessential to who they are that my Mongolian hosts can no longer recount the steps that led toward acquiring it.

Today's ignorance about the sky throughout much of the rest of the world was not always the status quo. While we like to think of ourselves, the present generation of humans, as well educated compared to our ancestors, there is much that we are becoming less and less knowledgeable about. Note that I am not thinking about obsolete technology, say, how to make the best whale-oil lamp; I mean fundamental knowledge about our environment. I am pondering the understanding of our own sky: what we see there, how what we see changes, and how it affects us.

I am interested in the astronomy accessible to everybody, without sophisticated instruments. It is the astronomy of the naked eye; you do not need a telescope to engage in it. (Everything described in this book I have seen with my own eyes.)

I do not refer to any sort of secret knowledge from the past. That is largely the stuff of graphic novels. Since people have become literate, we have been able to archive information. This assures that what we have learned about the sky is preserved. Moreover, thanks to modern science, there are specialists who know more about the heavens than our predecessors even dreamed of learning. Still, I maintain that the corporate sky knowledge of the rest of us is less than that of the average Roman citizen two thousand years ago.

There are obvious reasons for this. For the aforementioned Roman, astronomy was practical. It was good for business. If you were in charge of shipping grain from Rome to Carthage, then (as now) it would be important to know two things: when the deal was to happen and where Carthage is. Before clocks, timekeeping was done almost exclusively by watching changes in the sky. Even more recently, the sky was a vital source of data for mariners navigating the seas. (For some, it still is.) The information provided us by the sky is so important to modern society that we have replaced many functions of the skies with technologies more under our control.

At the same time, the sky is less accessible to us. This may sound strange. However, the average shepherd had a much better view of the sky than us. Not only was his horizon unimpeded by construction, but the darkness of the sky was total in the absence of artificial illumination. More important, that shepherd simply spent more time under the night sky. He did not share our indoor lifestyle. Plus there was little else to do during the dark of night besides look up at the stars. (TV reception was poor in the year one.) Our shepherd of the past, as intelligent as we are today, paid good attention to what he saw above. Eventually he recognized patterns. He and his brethren were among the first astronomers.

I will go so far as to say that astronomy was the first science. This sounds jingoistic, considering that I am an astronomer. But hear me out: I think this because astronomy began as the *simplest* science.

Do I mean that astrophysics is somehow simpler than, say, microbiology? No. However, science is an act of formulating theories. Two things are of great help in making the observations necessary to create predictive theories. These are repetition and regularity. The repetitions and regularity of the lights in the sky made it easier for our forebears to construct theories about them than did the more complicated patterns exhibited among, for example, flora and fauna (or other more complex phenomena of our terrestrial world). And so science was launched.

Today's professional astronomers have transcended simple observation of the sky with the unaided eye. They are interested in the physical nature of the universe and the forces at work there. They have been enormously successful at their work. Our standard textbooks now are filled with insight about distant worlds and galaxies. To make room for this new knowledge, chapters about what the sky actually looks like from the earth have been abridged. (This phenomenon parallels the decline in physical geography content within the social science curriculum.) What a nineteenth-century student might have spent a semester studying, today's counterpart breezes through in a few weeks. In the century before last, we did not even know that we lived in a galaxy of stars. Yet we were much more familiar with the sky and its cycles.

Those days of old are gone. We are so removed from the sky, and other realms of nature, that often we are not cognizant of ways in which they still affect our lives. Speaking more broadly, a lot of our culture continues to draw upon the sky by way of language, myth, and metaphor. To understand ourselves, I believe that the sky still matters. For when early people looked up at, and thought about, the sky, they really were trying to answer what is perhaps the most human question of all: Where am I? What is my place in the universe? So are we.

Chapter 1 discusses the astronomical sky in and of itself. What is its apparent shape and extent? How do we describe its population of stars?

In chapter 2 we develop a means by which we can communicate with one another about locations and apparent motions in the sky. We follow the seemingly moving stars in the sky throughout the night. Our reference is the horizon, where the earth and sky meet.

Chapter 3 introduces the concept of the celestial sphere to our cast of characters. The idea of a celestial sphere makes it easier for us to describe—not just *our* sky—but the sky as it is at different places on the earth. Also, we address changes in the sky that take place, not just over one night, but over the course of the year.

Long-term changes in the appearance of the stars are the subject of chapter 4. We conclude our examination of the stars with a discussion of how to get the most out of our human vision while sky watching.

From the stars we move on to the sun in chapter 5. The apparent motion of the sun in the sky is our principal timekeeper.

Moreover, the sun presents us with the gift of seasons, the topic of chapter 6. By understanding the annual apparent behavior of the sun, we humans have maximized its use in agriculture and for light and warmth. In the process we created civilization.

In chapter 7 we take a trip around the world to see how the seasons affect different peoples. We introduce "heliacal rising," historically the most important sky phenomenon that you have never heard of!

Chapter 8 treats the moon. We mark our calendar by the moon. Calendars are everywhere, and they are not always the same.

Analogies of the moon's apparent motion with that of the sun are explored in chapter 9.

Yes, but what do the sun and moon really look like? In chapter 10 we step back and examine the features visible on these famous disks.

Chapter 11 is about eclipses. We need both the sun and the moon to produce this comparatively rare phenomenon. At the same time, eclipses are so exciting that they deserve a chapter all to themselves.

In chapter 12 we "put" the planets in the sky, thereby completing our list of the objects regularly seen there. However, we also add a few ephemeral sights, to conclude the book.

(All descriptions of the sky herein are geocentric. They are based on an imaginary stationary earth. Until five hundred years ago, everybody assumed that geocentricism conformed to reality. We now know that the solar system is heliocentric—nearly centered upon the sun. I will use a bit of heliocentric jargon when discussing the sun; I think that it will be easy to tell which "centrism" I am using at any given time.)

1

Bowl of Night

And the first movement in the morning was to
open the window—again to examine the sky. . . .
I discovered a star—a solitary star—twinkling
dimly in the sky which had now changed its
hue to a pale grayish twilight, while vivid
touches of coloring were beginning to flush
the eastern sky. There was absolutely no other
object visible in the heavens—cloud there
was none, not even the slightest vapor. That
lonely star excited a vivid interest in my mind.
I continued at the window gazing, and losing
myself in a sort of day-dream. That star was
a heavenly body, it was known to be a planet,
and my mind was filling itself with images
of planets and suns. My brain was confusing itself with vague ideas of magnitude and
distance, and of the time required by light to
pierce the apparent illimitable void that lay
between us—of the beings who might inhabit
an orb like that, with life, feeling, spirit, and
aspirations like my own.

JAMES FENIMORE COOPER, "The Eclipse,"
unpublished manuscript circa 1831

This is a book about astronomy without optical instruments, the way
astronomy was practiced throughout most of human history. I will introduce some simple tools that people have used when studying the sky.
However, an instrument, narrowly defined, is a device that extends human
senses (a lens, a microphone, etc.). Here we will be aided by our eyes and
minds alone.

Before discussing what is *in* the sky, let us begin with the sky itself. We experience the sky as an imaginary inverted bowl, the center of which is directly over our heads. Its rim is defined by the horizon. When asked to look straight up, many of us do not bend our necks back a full ninety degrees. This and other psychological factors give us the impression that the sky is a somewhat out-of-hemispherical bowl, squashed at the top.

The sky-as-bowl (maybe a shallow bowl) is a practical way of thinking about the sky; after all, it really looks like that. And it is somehow comforting. Look up at the night sky, and try to force your eyes to see it for what it is: a black expanse as near to infinity as you ever will see, punctuated by stars at so vastly different distances from you that terrifically luminous stars appear dim and barely luminous stars appear bright. What did you feel in that instant before your brain again overtook your imagination? A bit of panic? Maybe a tinge of vertigo?

Thinking of the sky in "3-D" this way is difficult; it is contrary to any other human experience (where our environment is bounded). It can be exhilarating. It also can be scary. Our minds tell us that a bowl is a much more comfortable construct. There is nothing wrong with that. The idea of a black opaque bowl over our head is a useful fiction for describing things in the sky at any moment in time.

For instance, it aids communication. You can tell another person where to look at some shared sight in the sky by providing him or her with just two pieces of information. One is how high to look, the other is where (in which direction) to look.

By "height" and "where" I do not refer to *physical* distance. I just mean how far one must tip one's head, and how to turn, to see the star in question. (I will quantify this at the beginning of the next chapter.) Without prior knowledge, it is impossible to know how far away a star—or anything, for that matter—is. After all, once an object is too high to reach (e.g., an apple in a tree), would you know how far away it is without first knowing how big an apple is? Likewise, how far are the stars if you do not know how big a star is? You have never picked up and examined a star. You have read in textbooks that a star is exceedingly large, but would you necessarily assume that without being told?

On the earth we may get lucky: There are clues to distance that are familiar to artists who deal in perspective. A landscape feature nearly lost in the haze is likely farther away than a similar one in clear view. But we have none of these clues in the sky.

The bowl of the sky has been an important philosophical concept in

Western society. For instance, in the Bible, it is called the Heavenly Firmament. It served to separate the earth from the heavenly waters. It is not outrageous to think that there is water above us: What color is the sea? Blue. And, of course, water does sometimes fall to the earth—in the form of rain.

(In popular culture, the idea of the sky as a physical thing has always had great traction. "It's like the sky is falling" remains a simile for calamity.)

Unfortunately, the view that the heavens are separate and different from the earth obscured the fact that we have a great deal in common with the rest of the universe. Indeed, we are intimately related to it. The suggestion that the "rules" are different "up there" and "down here" retarded Western science for millennia.

Even in more modern times, since the idea of a physical star bowl was discarded, a sophisticated argument reinforces the sky-as-bowl metaphor. A bowl has a finite thickness.

Consider that most of the night sky is black. There is no star in most random directions. These two statements may seem trivial, but they are far from it.

Why are there not stars everywhere in the sky? A literal bowl of stars would seem to save us from this question. Its thickness is presumed to be small compared to its radius. The number of stars and the volume of the bowl have a fixed ratio that allows for many directions in which we look and encounter no star at all.

Yet as soon as you allow stars to be at greatly different distances away, and no longer require them to be embedded in a physical matrix, there is the possibility that the *space* of stars might extend forever. If so, every line of sight eventually must run into a star. Starlight dims but does not disappear with distance. An infinite space logically implies a sky completely ablaze with stars. This is not the case. So some sort of limit to the distances of the stars is sensible: Our metaphorical bowl of sky has a certain thickness. Notice that this argument works regardless if you believe that the universe is centered upon us, or whether you have made the Copernican leap in which the earth is not necessarily the center of everything.

After wrestling for hundreds of years with the paradox of a sky not infinitely populated with stars, astronomers finally did away with the requirement of celestial boundaries. We now know that 1) starlight is not instantaneous—light travels at a constant velocity, and 2) that we inhabit a universe of finite age. There is a distance so far away from us that we cannot see starlight originating at that distance. The travel time for this light would exceed the lifetime of the universe. Far from an elementary matter, a star-speckled (but not blanketed) sky sends a profound message to us about the nature of our universe.

THE ARCHITECTURE-SKY CONNECTION

When nearly everything else a society produces has disappeared, its buildings usually are what remain. This is why I will write about a lot of buildings and other monuments in my attempt to convince you that people have interacted with the sky for a long time. The study of prehistoric peoples' interaction with the sky of sun, moon, planets, stars, and so on, is called archaeoastronomy.

I will use two words in this book that at first sound the same; however, my meaning for each will differ. One is "orientation." An orientation occurs when two or more objects (or a plane) point in a particular direction. Orientation does not require intent. On my campus, two Spanish cannons are oriented with (aimed at!) the Methodist student center across the street. Nobody seriously thinks that this was intentional. It just happened. I will use the more subtle word "alignment" for an intended orientation. Warning: Some scholars juxtapose the definitions of these two words. However, to my ear, this usage sounds best.

An alignment or orientation requires both a backsight and a foresight. A sphere is worthless insofar as determining a preferred direction. We need an asymmetrical object with at least two discernable points (ends, edges, openings). The one closest to us is called the backsight, the one closest to the point to which we are to look is called the foresight. The notches on a rifle are effective back- and foresights.

The earliest purported use of astronomical foresights and backsights of which I know is at the Nabta Playa in Egypt.[1] Here, sandstone blocks were moved into apparently intentional celestial alignments, by pastoral peoples as early as the fifth millennium BCE![2]

It used to be that to measure the orientation of architecture, one had to become familiar with the ways and means of the surveyor. Both direction and inclination must be taken into account. (A hillside has a different natural horizon than does a flat plain.) Today, though, the marvelous invention of GPS technology may greatly simplify the task.

Another example of archaeoastronomy can be found in the middle of America. About the year 1200, the Mississippian civilization built a city called Cahokia across the river from what would become modern-day Saint Louis. You can still visit it: It is on the United Nations list of World Heritage

The Darkness of Night

Look down into the abysmal distances!—attempt to force the gaze down the multitudinous vistas of the stars, as we sweep slowly through them thus and thus—and thus!

Even the spiritual vision, is it not at all points arrested by the continuous golden walls of the universe?—the walls of the myriads of the shining bodies that mere number has appeared to blend into unity?

EDGAR ALAN POE, "The Power of Words," 1845

The question, "Why is the sky dark at night?" often is called Olbers's paradox, after the eighteenth-century astronomer Heinrich Olbers. In recent times it was popularized by oft-read cosmologist Hermann Bondi (1919–2005). Astrophysicist Edward Harrison (1919–2007) traced its history in his book *Darkness at Night: A Riddle of the Universe* (1987).

It is an oddity of science that ideas and phenomena often are named after the wrong person. The darkness-of-night question was stated correctly, in print, by the French aristocrat Jean-Philippe Loys de Chéseaux (1718–51). (It predates him, though.) Chéseaux answered his own question by saying that a not-quite-transparent medium (gas?) between the stars absorbs the light from the most distant stars before it reaches us. Modern physics points out that this is, in fact, no solution: All the energy from the myriad of stars cannot be lost at absorption—the absorbing medium must eventually heat up to the point at which it begins to glow itself.

Heinrich Olbers (1758–1840) was a famous German astronomer who discovered two of the largest minor planets. He proposed the paradox in 1823, in much the same way (and with the same incorrect deduction as the solution) as had Chéseaux in his 1744 publication *Treatise on the Comet of December 1743 to March 1744.*

Olbers owned Chéseaux's nearly eight-year-old book. Presumably he read it. Let us be charitable and assume that Olbers forgot about Chéseaux's work and actually thought that the paradox was original with him. Regardless, the name of Heinrich Olbers is better known in astronomy than that of Jean-Philippe Loys de Chéseaux. (Chéseaux's early death did not help his fame.) The question in its current form was credited to Olbers by the great textbook author of the time, John Herschel (1792–1871). And once it made it into the textbooks, it was assured that "Chéseaux's paradox" would always be Olbers's paradox.

The unambiguous, correct solution to Olbers's paradox was posited by popular-writer Edward d'Albe in his 1907 book *Two New Worlds*. However, in that Edwardian age, when nobody had ever so much as heard of a "big bang theory," the idea of the universe's finite age was dismissed for what it was: d'Albe's speculation.

Sites. Evidence of the Cahokian people's existence lies underground in excavated artifacts, and above ground in the world's largest platform mounds made of earth.

West of the major Cahokian mound, one on which the supreme ruler of Cahokia is hypothesized to have lived, there was once erected a seventy-meter-diameter circle of tree trunks—forty-eight in all. In the middle was a lone post. Standing at the interior post (backsight), it is possible to see the point on the horizon where certain objects rise and set in the sky at particular times of year, behind three of the outer poles (foresights). I will discuss the solstices and equinoxes later. Right now, it is important to realize that what I have described so far is an orientation, but not an alignment.

With so many posts in the Cahokia sun circle, it is possible that these events occur by chance. We need further evidence to suggest an alignment. This evidence comes in many forms—from archaeology, other sciences, or somewhere else—and is somewhat subjective.

However, it is necessary to distinguish human accomplishments that are alignments from all the orientations that *must* (statistically) occur.

It is easy to skip this warning about mistaking orientations for alignments because our minds are designed to 1) pick out patterns and 2) pick up on exceptions to normal life: We notice when we see "12:12" light up on the digital alarm clock, forgetting all the times that we have glanced at the same clock seeing no such interesting pattern. We remark on the coincidence, even though something like "12:12" must show up during one minute of every hour of every day.

At Cahokia, we suspect a real alignment because the observed orientations do not work if the middle pole is at the geometrical center of the circle—as it likely would be if somebody was merely building a "pretty" wheel of wood poles. Here we use the scientific method: We hypothesize that the sun circle is meant to mark natural phenomena; therefore, we figure out where one must stand for the alignment to work. This turns out to be a couple of meters east of the geometrical center. We then test our hypothesis by digging for the remains of an ancient post at this spot. It is there. (Notice that we did not dig a hole everywhere we could within the sun circle, hoping to find something; we dug in one—and only one— predicted location.) It looks like the builders of the Cahokia sun circle ignored the strong human urge for symmetry (placing the middle pole equally distant from the perimeter poles) and put function over form: The Cahokia sun circle may have looked a bit off, but it worked.

Is this enough evidence? Now the archaeologists weigh in. There seems to be a connection between one of the outer poles and the sun.

The Dawn of Archaeoastronomy

The founder of modern archaeoastronomy has to be Sir Norman Lockyer (1836–1920), Fellow of the Royal Society and an astronomer famous for the discovery of the element helium in the sun. He was able to make this discovery, before helium ever had been isolated on the earth, using the new technique of astronomical spectroscopy. He also established one of the most prestigious scientific journals in the world, *Nature*.

While on holiday, later in life, Lockyer became fascinated with Greek and Egyptian temples and monuments. He began to make systematic measurements of, and calculations based upon, their orientations (though not without mistakes). Lockyer continued to use his spare time in order to "prove" that these buildings were aligned to the sun and stars. His 1894 publication of *The Dawn of Astronomy* can be considered the birth of the interdisciplinary science archaeoastronomy.

However, Lockyer turned out to be a problematic founding father. Archaeologists were appalled by his book. It was clear by his writing that Lockyer did not know much about his subjects, and that when he did not know something, he replaced fact with broad speculation. Lockyer (and other astronomers) complained that the archaeologists did not know basic astronomy, a claim that was probably equally true. Later, Lockyer turned his attention to his homeland and wrote *Stonehenge and Other British Stone Monuments Astronomically Considered* (1906). Many of Lockyer's assertions turned out to be false. A few turned out to be true.

Norman Lockyer founded a discipline. Yet, in doing so he created a schism between archaeologists and astronomers that is only now healing. Because a transdisciplinary approach is vital to successful archaeoastronomy, Lockyer and his methodology greatly impeded the new discipline's progress, too.

They dig there. The excavators find what they know (because of their study of Cahokian culture through a myriad of other artifacts) to be a small ceramic vessel, the kind used to hold religious offerings, with a symbol representing the sun adorning it.

Do we have enough evidence now? We can never prove as a mathematical certainty that the Cahokian sun circle was aligned based on the astronomical knowledge and interest of its people. Questions remain: What were the other poles for, then? Still, there is enough evidence to make it unlikely that it was not aligned.

STARS AND CONSTELLATIONS

It is time to populate our sky. When we look up at the night sky, the moon may or may not be present; the same is true for the planets. We *might* see a meteor, comet, or some other rare astronomical phenomenon. However, we are guaranteed to see the stars on a clear night. Under optimal conditions, the naked eye may perceive thousands of stars above the horizon at one time. English amateur astronomer Thomas Backhouse (1842–1920) published a book with the title *Catalog of 9,842 Stars Visible to the Naked Eye* (1911).

Sky watchers group the stars together into patterns called constellations. This is an old human habit. Some say that certain constellation patterns are tens of thousands of years old and survive until today largely by word of mouth, passed from generation to generation.

Constellations are patterns only, kind of a celestial connect-the-dots game. They have no physical significance. Some stars in a given constellation may be vastly farther away than others in the pattern. Thus, viewed from anywhere other than our solar system, the pattern would change. (Have you ever seen a rock or mountain with a profile that reminded you of a familiar shape, perhaps a human face? The particular pattern disappears when you step away from your viewing spot.) Still, constellations are handy landmarks for someone standing on the earth.

At one time constellations were named after creatures, heroes, and stories from popular mythology. Only the brighter stars were used. Some of the figures are obvious even today, though most take a great deal of imagination. Yet this was not a game such as that played by a child lying on her back and identifying elephants and mountains among the clouds. Remember that the creatures, heroes, and stories existed first. People assigned star patterns to honor these elements in their culture, not necessarily because the pattern looked that much like that for which it was named. (Someone once said that the constellations need look like their namesakes as much as the state of Washington resembles the United States' first president.)

The constellations as established were more-or-less "stick figures," with no agreed-upon boundaries. Moreover and somewhat obviously, different civilizations recognized different sets (and names) of constellations.

Modern astronomers have formalized the constellations, fixed their number (eighty-eight), and assigned them definite boundaries. The result is that every star is now a member of some particular constellation. You often can recognize the new constellations: For example, if you come across Antila (the Air Pump), you rightly will suspect that it is not of ancient origin. Smaller constellations may be so because they had to be wedged in between older, larger constellations. My least favorite constellation is Triangulum. Inasmuch as any three stars in the sky will form a triangle, I find this name singularly unimaginative.

A map of the constellations today looks a bit like that of the western United States: lots of east-west and north-south running borders. Now constellations can be used to give a general description of a location in the sky without resorting to a numerical coordinate system. Often this is all that is needed. If you are driving to a new city, you may not require its exact latitude and longitude on the globe. Knowing what county it is in might be all that you need to navigate your way there. If the city is big enough, you will see it once you get close.

Knowing what constellation an astronomical object occupies may be all that you need to find it, if the object is bright. The only requirement is a chart of the constellations.

Of course, you also need to know which constellations will be visible in the night sky at specific times, in order to know when and where to look for the object. Soon I will introduce this time element into our discussion.

BRIGHTNESS AND COLOR

Besides their apparent position with respect to each other, do the stars all pretty much look the same? To a first approximation, they do. Stars are points of light in the sky. The "starburst" pattern of rays we might think we see emanating from a bright star is actually due to an optical effect within our own eyes.

For centuries we have wrestled with the apparent size of stars. Thinking that stars must have *some* apparent diameter, people equated brightness with size. This impression was proved wrong when the telescope was invented and revealed that, no matter how much a star is magnified, it remains point-like. Stars are so far away that we cannot see their apparent size without some physics tricks available only in the last century.

The word "stellar" is an adjective meaning "of stars." The fact that stellar brightness does not correlate with size—that stars do not have apparent size at all—came as a shock. It *looks* like the brightest stars are tiny disks. However, this is an optical illusion. Stare at those stars: Do they really have a shape? Does that shape remain constant? No. Logically, something with an apparent size must have some constant shape. The twinkling stars do not.

But stars do differ. It is readily obvious that the stars vary in brightness among themselves. This is due to their varying distances from us and their different luminosities. What is the difference between these two words? A floor lamp appears to differ in brightness, depending upon how far away you stand from it. However, the bulb in the lamp has a constant luminosity—the intrinsic amount of light emanating from it. (The luminosity of an incandescent light bulb usually is stamped on it, e.g., "60 watts.") As there is no way to separate the effect of distance and luminosity in stars just by looking at them in the sky, hereafter I will use the descriptive term "brightness" exclusively.

Interestingly, the difference in brightness among the stars in the sky is greater than that admitted by our eyes. When the human eye and brain estimate that one star is twice as bright as another, in reality, they differ by a factor closer to 2.5 (as measured objectively by electronic instruments). The range in brightness between the brightest stars and those we can just barely see under optimum conditions is roughly one hundred. The range from invisibility to Sirius, the brightest star in the sky, is even greater.

Astronomers have placed the stars into brightness categories called magnitudes. First magnitude stars are the brightest; sixth magnitude stars are only just visible. Each change of one magnitude represents a two-and-a-half-times change in brightness. There are about

20 stars of magnitude 1 (or brighter);
65 stars of magnitude 2;
200 stars of magnitude 3;
450 stars of magnitude 4;
1,100 stars of magnitude 5;
4,000 stars of magnitude 6.

Not everybody can see a sixth magnitude star. (Furthermore, we will discover that not all the stars listed above are visible at the same time.) On the other hand, there are people with visual acuity such that they sometimes can see thousands of additional, fainter stars than those listed above.

Two stars of similar magnitude still need not look the same. Stars of different temperatures have different colors. However, do not be too disappointed if you do not see any stellar color at all. Individuals' color perception varies markedly. Some people do not possess all three color receptors in their retinas; they are at least partially colorblind.

Also, astronomers speak of red and blue stars with an acknowledged exaggeration. The variation in color between the most "red" of stars and the most "blue" is less than that between the different varieties of street lighting, all of which you may consider to be white. Star colors are desaturated—they contain a great deal of white mixed in with pure color. Plus, they are point sources, which our brain tends to register as whiter than extended sources of the same chromaticity. In truth, only about 150 (bright) stars may be said to be "colorful" as seen by the naked eye.[3]

The brighter the star, the easier it is to perceive its color. The eye needs a certain amount of light to trigger color vision. This is why, when we awake in a dimly lit room, everything around us appears in shades of gray.

The bright stars Arcturus and Antares normally are seen as red. Procyon and Capella usually are considered to be yellow. Some say that Castor is green-tinted. Vega approaches blue.

If you do see a colored star, it is likely to be red. This is because there are more particularly bright red stars in the sky than there are blue. This is not a coincidence. Astronomers have learned that red stars can be enormous, and correspondingly luminous. So, assuming stars are scattered at random distances from us (a crude but useable approximation), it is likely that some of these big "high wattage" stars are near to us and appear quite bright. They are and do.

BEYOND THE NEARBY STARS

Besides individual stars scattered about the sky, there is another, distinct population of stars always visible on a clear night. However, you may not have recognized it as such. In reality, the sun and its planets are located in a galaxy of stars that is disk-shaped. We are situated in the plane of that disk. So in a particular set of directions, there are more stars than in any other (those pointing within the plane of the galaxy). Imagine standing in a crowd of people roughly your own height. As you turn your head every which way, most of the time you see no faces. You see sky. You see shoes. But not faces. Yet in one plane, the plane five-feet, nine-inches above the floor, you see lots of faces.

Many of the stars in the galactic plane are extremely far away. Still, there are so many stars in this set of directions across the sky that their combined light blends together. From an exceptionally dark vantage point, we see a seemingly continuous band named the *Via Lactea*, or Milky Way. The ancient Chinese called it the Yellow Road, and in John Milton's *Paradise Lost* (1667) it is the Road Whose Dust Is Gold.[4] These latter names suggest a color that most of us cannot detect with the naked eye.

The Milky Way is slightly brighter in the direction of the constellation Sagittarius. This is the direction toward the center of the Milky Way galaxy. There are approximately ten times more stars visible (per unit area of sky) in this part of the Milky Way than in the portion found in the direction of the constellation Auriga.

It turns out that people who live in the Southern Hemisphere of the earth get a better view of the Milky Way than do those who live in the Northern Hemisphere. (I vividly recall seeing the famous statues of Easter Island clearly silhouetted by the background light of the Milky Way alone.) Southerners have gone so far as to use the orientation of the Milky Way—sometimes it may lay along the horizon, other times it may pass directly overhead—to keep track of time.

The Aboriginal Australians made their home on the coasts, but also the Great Outback, a place of severe climate but breathtakingly dark skies. They noticed dark patches *within* the Milky Way, which we now know to be caused by interstellar dust clouds blocking our view of more distant Milky Way stars. (The most obvious of these apparent "gaps" in the Milky Way is the Great Rift, between the constellations Cygnus and Scorpius.) People all over the world have made up constellations from star patterns in the sky. The Aborigines invented constellations out of the patterns formed by these *absences* of stars.[5]

(Interstellar clouds of gas and dust in our Milky Way galaxy need not be dark. Occasionally they are made brighter than the black background of space by reflecting starlight. Alternately they heat up, due to the presence of nearby stars, and become self-luminous. It is often amateur astronomers who argue over which of these nebulae are visible to the naked eye. I find this debate to be of little interest, because in order to spot a nebula without a telescope, it is necessary to know exactly where to look—and this generally requires a telescope.)

Even the stars of the Milky Way galaxy are not the most distant we can see without optical aid. From a very dark location, on a moonless night, look in the direction of the constellation Andromeda. If you have a

FIGURE 1.1.   Here a dark campsite in the Cascade Mountains is photographed using a fisheye lens. The horizon lies all around the border of the picture. The Milky Way extends from horizon to horizon, across the middle of the sky. (In choosing images for this book, I sought those that realistically depict what you might be able to see with your own eyes. Exceptions are noted in the captions.) Photo by Rik Littlefield.

keen eye, you may see a tiny smudge of light, not point-like as you would expect from a single star. This is, in fact, another galaxy of stars. It is the Great Andromeda galaxy, which may be even larger than our own. It is far, far away from the Milky Way galaxy. The light from hundreds of billions of stars combines to make the Andromeda galaxy the most distant object visible to the unaided eye.

What can we say about the entire universe itself? To be sure, only a tiny fraction of it is available to the human eye alone, limited as that organ is by finite brightness sensitivity and restricted as it is to a narrow range of the electromagnetic spectrum. (We cannot see heat, we cannot see radio, and we cannot see gamma rays: The universe emits all of these, and more, besides visible light.)

Quite clearly, the observable universe is profoundly dark. The light we see in the sky is exceptional. On a lark, two astronomers once set about averaging what color the observable universe is. Their answer? A pale green.[6] But nobody takes this "answer" seriously. We must acknowledge that our earthbound view of the universe is not the whole picture. Yet what a lovely picture it is.

MAKING A POINT

It is easy to tell somebody about the stars. It is not so easy to show them the stars.

Imagine that we are outdoors and I am pointing out the locations in the sky of some of the stars, and so on, that I write about. Or you are pointing them out to others. How to do it? To make sure that another person actually is looking in the same direction that you are pointing your finger, it is necessary to be behind him or her—looking over the person's shoulder.

A battery-operated laser pointer makes the job easier. This small device produces a fine beam of light that does not spread out (much) with distance. It illuminates dust and aerosols in the air on its way upward, forming an arrow to whatever it is you are pointing it toward. Groups of people can see at the same time where it is you are indicating in the sky.

I do not mean the small, red pointers used on a lecture-room screen. You need something brighter than this outdoors. I mean the higher-power green-light lasers. Be careful, though. These tools must be used with care. A thoughtlessly pointed laser beam that ends up in someone's eye can cause serious harm. This is still true at great distances away, because the beam attenuates so little. Never point a laser upward when there is an airplane in the sky.

My Celestial Desiderata

My first "grown-up" astronomy book was *A Field Guide to the Stars and Planets,* then edited by astronomer Donald H. Menzel. (I still possess that worn copy.) It was and is part of Houghton Mifflin's time-honored Peterson Field Guide series. Amazingly, this book remains in print, and still is incredibly useful. It now is edited by my colleague Jay Pasachoff (1999).

My single favorite astronomical almanac (astronomical information for a given year), pertaining to the night sky, is the *Observer's Handbook,* published annually by the Royal Astronomical Society of Canada (Toronto). It presently is edited by Patrick Kelly. Of particular interest to us is the section "The Sky Month by Month."

No product endorsement is intended, but I also enjoy hanging Celestial Product's MoonLight on my wall each January. This chart is in reality a graph depicting the phase (and therefore the brightness) of the moon for each day of the current calendar year. (See chapter 8.) The moon blinds: The more moonlight, the less starlight visible in the sky.

It is an ironic fact of meteorology that, especially in the winter, the clear nights are the coldest. What works in the way of apparel for running from heated home to heated car to heated building will not do for standing still outside, looking up. A professional observer named Scott Murrell, a veteran of decades in an unheated telescope dome, once told me that his most important piece of "equipment" was his Duo Fold thermal underwear! A warm hat and boots (at the least) are important astronomical accessories, too.

Also, do not do what I did when I was a kid: I borrowed my school's laboratory laser (a much bulkier, much less powerful predecessor of the laser pointer) and pointed it at the sun sensor atop the streetlamp outside my house. When I wanted to "do astronomy," I fired the laser, causing the lamp to react as if the sun had risen and shut it off. The sky was dark again; gone was the annoying artificial light. Think of the liability to which I exposed my parents.

2

This Big Ol' Wheel
Keeps Rolling

If the eyes do not look at the Sky,
what else would they look at?

IGBO PROVERB[1]

When we talk about how "high" something is in the sky, astronomers refer to an angle. There is no spatial measurement that one can make in the sky without knowing about the physical nature of the objects in the sky. And the night sky by itself is secretive about revealing that information.

ALTITUDE AND AZIMUTH

How can you tell somebody where to look in the sky? You can point "over here" or "over there." But what if you cannot use your hands—what if you wish to communicate with somebody not physically present? By words alone?

Astronomers define the angle between an object in the sky and the horizon as its altitude. This is a different use of the word than the more common one applied by, for instance, airline pilots. We have heard them over the loudspeaker saying, "We are climbing to an altitude of thirty thousand feet." No, the astronomers' altitude is an angle, measured in degrees, with zero at the horizon and ninety directly overhead.

Admittedly, our horizon is often interrupted by trees, buildings, hills, and McDonald's restaurant signs. However, right now we are discussing the real astronomical horizon, the one that would exist if you stood on a featureless plain.

Here is a simple and approximate way of measuring altitude in degrees: Extend your hand vertically at arm's length from your face. If your

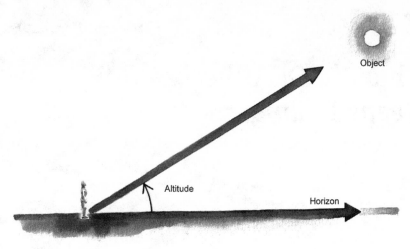

FIGURE 2.1. Measuring astronomical altitude. The observer is at the intersection of the two imaginary rays.

fingers are spread apart, they extend about twenty degrees in the sky. A closed fist covers about ten degrees. A fingernail, one degree.

Your hand is not the size of a constellation. Your arm length is not the distance to a star. However, due to the geometric property of similar triangles, this method works for measuring altitude.

If this sounds bogus, it may be because you realize that all people are not created alike. Do not some people have long arms and some short? Do not some have small hands and others large? All true. But typically, the long-armed people also have the larger hands—and vice versa. The result is that the ratio of hand size to arm length is more or less constant for everybody. So the angle measured in this way is more or less the same from person to person.

Chinese navigators once carried with them wood squares of differing sizes. Holding a square up at arm's length represented a predetermined angle in the sky. While a slightly more objective measurement than that described above, this system could be made even better by customizing the squares to the navigator.

An alternate method for measuring sky angles is with a cross staff.[2] A short (usually) wooden bar slides along another, longer wooden track at right angles to it. The user looks down the track at the bar, sliding the bar until its apparent angular extent overlaps that between the objects (or horizon and object) being measured. The angle is then read off a scale on the track.

Sailors of the Indian Ocean used to carry with them a kamal. This was a piece of wood with a hole in the center. Through the hole was threaded a string. While holding the string by mouth, the user could slide the wood piece along the taut string. Its apparent size changed. The navigator moved the wood piece so that it covered an angle of interest. Perhaps a star was at the top of the wood piece, and the horizon was at the bottom. Knots in the string marked locations where the wood piece took up a given angle. The number of knots passed by the wood piece indicated the measure of the angle.

Notice that these latter two methods of measuring sky angles are identical in result to the other methods described. Instead of changing the physical size of the measuring tool (hand posture, wood block) and leaving the distance to it (an arm's length) constant, this technique leaves the physical size of the measuring tool (cross-staff, kamal) constant, and varies the distance to it. Generalizing, if you know two of the following pieces of information—

1. the physical size of an object (or physical distance between two objects);
2. the physical distance to the object (or objects);
3. the apparent angular size of the object (or apparent angular distance between two objects)

—you also know the third.

Besides altitude, the other piece of (angular) information necessary to convey, so that you and a partner are looking at the same thing in the sky, is called azimuth. By itself, altitude tells your counterpart whether to look near or far from the horizon—but in which direction? Azimuth is a more familiar concept than altitude. It is an angular equivalent to compass direction.

We are used to giving directions here on the earth. We tell somebody to look north, south, east, or west. Or perhaps northeast, southwest, southeast, northeast, and so on. The azimuth angle is measured starting due north, and then increases as we turn (eastward) to 180 degrees at due south and 360 degrees by the time we are looking due north again.

Those numerical signs on airport tarmacs tell pilots the azimuths of runways. The signage has just dropped the last digit. An "18" runway points due south (180 degrees in azimuth). Approaching the same runway from the other end, the pilot will see "00." It is, of course, also a north-pointing runway.

Any of the methods I have discussed for measuring altitude work for measuring azimuth, too. A great circle on the sky of constant azimuth (plus or minus 180 degrees) is called a secondary or vertical circle.

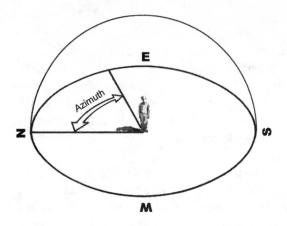

FIGURE 2.2. An azimuth of eighty degrees is a bit north of east.

That is it. Two angles define a unique spot on the bowl of sky. They are all that we need to navigate our way among the stars in the sky—at least until we notice things moving.

RISING AND SETTING

It takes only a couple of hours, perhaps less, to notice that the sky changes. A hypothetical celestial object completing a circle centered upon us will appear to travel 360 degrees in just about twenty-four hours. There is one circuit per day. Two hours is to twenty-four hours as fifteen degrees is to 360 degrees. Fifteen degrees is easily measureable even with the hand-and-arm method.

Two hours or less. That is not much time, but often more time than modern humans are willing to give the sky as they exit an enclosed workplace, in an enclosed car, to drive to an enclosed house. When we are outside, it is often in the daytime, and often our attention is directed to something down: the sidewalk, the lawn mower, the golf ball. When we are indoors, we live in a virtual world without sky.

For these reasons, especially city dwellers often do not know that the stars appear to move each night, or if we admit motion, that the way in which the stars move is not haphazard. This is no fault of the urbanites: In the city, where the sky is defined as a narrow slit between high-rise buildings, it is even understandable.

Regardless, the stars do appear to move, and in an orderly way. For instance, simplifying things significantly is the fact that the stars all seem to move together, as one unit, as if they were attached to a solid, moving background. They are not. Still, that is what it looks like.

The stars form the canvas on which any astronomical spectacle plays out. Stars are so distant that they physically move imperceptibly to

Why Are Astronomers Seemingly Obsessed with Geometry?

Astronomers use a lot of geometry because they have to. Once something is higher than we can reach, estimating physical size and distance becomes problematic. One thing we can do, though, is measure angles in the sky. So we speak of apparent size, measured in degrees, instead of physical length, height, or width, measured in meters, feet, or cubits. The hand method for estimating angles works for more than measuring altitude and azimuth. It can be used to measure the apparent angular size of something (e.g., a constellation) or the angular distance between two astronomical objects (say, stars). Moreover, it can be used to measure change of angular position in the sky over time. Degree measurement is convenient because most people remember that there are 360 degrees in a circle. In our case, the circle is one centered on the observer.

the naked eye, within the span of human history. The much closer planets, satellites, and minor bodies of the solar system, on the other hand, change their relative locations as they and the earth move about the sun. Thus, the stars, which appear fixed, serve as a ready reference for specifying the address of another astronomical object, independent of time.

Consider a series of imaginary stars (though they need not be imaginary). Here I choose a sky view from a site in the midlatitudes of the earth's Northern Hemisphere (where I happen to live). One star rises far to the southeast, increases in altitude as it follows a circular arc, and sets in the southwest. Its altitude at star rise and star set is, by definition, zero. It reached its maximum altitude halfway through the trip, when it was due south; even this maximum altitude was not great, and the star spent comparatively little time at all above the horizon.

Meanwhile, another star rises further north, though still in the southeast. This star also sets more northerly. It reaches a greater maximum altitude and stays above the horizon longer than our first star.

What about a star that happens to rise due east? There is no physical reason why there might be such a star; the stars do not arrange themselves based on earth geography. Still, there are a lot of stars. That one happens to rise due east and set due west is not farfetched. This star achieves a greater "peak" altitude (though still in the south) than either of the previous stars. It stays (almost exactly) twelve hours above the horizon and (almost exactly) twelve hours below.

A star rising in the northeast sets in the northwest. Because we are in the Northern Hemisphere of the earth, this star gets high in our sky—perhaps passing overhead—and spends more than twelve hours in view.

THE CARDINAL POINTS

By now you cannot help but notice the overall pattern of apparent motion exhibited by the stars. It all seems to turn about a point. To define a spin axis, one needs two points. So logically there must be another such point, too. However, it is always below our northern midlatitude horizon.

It is clear to any observer that the sky is not in all ways symmetric. There is a preferred axis north to south, the axis mundi (axis of the world). The existence of north and south invites also the existence of east and west. An imaginary east-west line, called the equinoctial, is always perpendicular to the axis of the earth and sky.

Cardinality is perhaps the most fundamental gift imposed by the sky upon the earthly inhabitants below. The concept of a four-quadrant

FIGURE 2.3.   Looking south. In this long-exposure photograph, the stars' light smears, tracing out segments of circles. Notice that the lights of a vehicle, on its way to the observatory, likewise make trails. Courtesy of Gemini Observatory/Association of Universities for Research in Astronomy.

FIGURE 2.4. The same photography trick used in figure 2.3 shows stars rising in the east. Courtesy of Caltech/Palomar Observatory.

grid superimposed over topography pervades a great many cultures, both old and new. In writing, it is as ancient as Homer, and as modern as Moby and Gwen Stefani: All four cardinal points get their due in the pop song "South Side," which peaked at number 14 in 2001.

The intercardinal points are northwest, northeast, southwest, and southeast. These can be further divided and divided again. The result is the thirty-two-point compass rose. For instance, there are eleven and a quarter degrees between north and north by east, eleven and a quarter degrees between north by east and north-northeast, eleven and a quarter degrees between north-northeast and northeast by north, and eleven and a quarter degrees between northeast by north and northeast. And that just takes us one-eighth of the away around the compass rose. As planetarium director George Lovi points out, though, filmmaker Alfred Hitchcock got it wrong: There is no such thing as *North by Northwest*.

At one time, cadets at the U.S. Naval Academy (and other maritime academies) were hazed by their upper classmen using the compass rose. It is called "boxing the compass": On demand, the new student had to recite the compass rose from memory, while standing at stiff, formal attention.[3]

A word about direction finding: The magnetic compass is a useful device—try finding north on a cloudy night without it! But magnetic north is not quite the same thing as true north. Moreover, the magnetic

Extinction: Star Rise & First Appearance Are Not Always The Same Thing

We observe a star through a blanket of atmosphere. While transparent for the most part, the earth's atmosphere does contain opaque materials (dust, ice crystals) that dim starlight. The more atmosphere we look through, the brighter a star must be in order for its light to remain perceptible to us. The atmosphere is a nuisance to astronomers—except for the fact that we need it to breathe.

Our planet's atmosphere is a layer of gas surrounding the solid globe. It is, effectively, only about one hundred kilometers thick. Thus, it is thin compared to the size of the earth. From our point of view, then, although our geography books teach us that the world is "round," we can picture ourselves as standing at the bottom of a flat dish of air. The dish is wider and longer than it is tall. We want to stare upward through as short a column of obscuring air as possible. This path is directly overhead.

Think of standing (or floating) at the bottom of a large swimming pool. Soon you run out of air to breathe—but you are at the middle of the swimming pool. What do you do? Do you swim for the nearest pool side? Likely not. You bob straight up. Intuitively, we know that in this direction we have the least water to pass through before surfacing.

Astronomers call the length of the air column they are looking through the air mass. (Directly overhead is defined as 1 air mass.) Observing a celestial object at any increasing angle with respect to the point directly overhead means that you are peering through a greater and greater air mass. Your view is dimmed. As you get closer to the horizon, the view deteriorates quickly. Indeed, it is feasible to see only the brightest astronomical objects (e.g., the sun and the moon) *at* the horizon.

All stars are extinguished from naked-eye view before reaching the astronomical horizon. Most of the time, as a star changes altitude, it also changes azimuth. Thus the azimuth at which a star disappears is not exactly the same as the azimuth at which it technically sets. The dimmer the star, the longer is the interval between extinction and setting—and the greater the difference in azimuth. The azimuth effect is more appreciable the closer you are to the earth's poles and the further away you are from the equator. Notice that, in the Northern Hemisphere, extinction produces a systematic error in apparent setting direction toward the south (and vice versa). Yet the exact extinction altitude depends on the local condition of the atmosphere night to night.

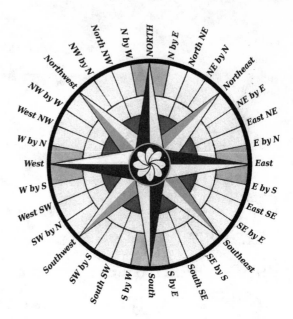

FIGURE 2.5. The compass rose sometimes is called a wind rose.

properties of the earth beneath and objects around you can affect the direction in which a compass needle points. Two compasses, used at two different places on the globe, may not indicate the exact same "north." When we talk about the sky, we refer to true north. For instance, when an astronomer says "south," he or she always means "in the direction of that big 'barber pole' explorers (with a certain sense of humor) erected on the earth's South Pole." *Geographic* south.

AN EXAMPLE OF CARDINAL ALIGNMENT

The human architecture based on cardinal alignment would exhaust cataloging. It includes my hometown of Cedar Falls, Iowa, where, disregarding the natural topography of the Cedar River basin, the numbered streets align east-west, and Main Street runs north-south. Cedar Falls sits in one of Iowa's ninety-nine counties. Iowa county lines form a grid work with east-west- and north-south-running borders. Iowa's northern and southern borders themselves both are aligned cardinally, as is the northern border of the United States. There is a continuous, cardinal progression on the map, from my exactly east-facing house to the axis mundi.

Let me share just one example of cardinal alignment from antiquity—a personal favorite that was constructed more than two millennia ago. Near Xian, China, is the tomb of Emperor Qin, the ruler who first united China, both politically and culturally, in the third century BCE. Qin's influence on Chinese history was profound. The Great Wall of China? Yep. That was Qin.

Qin's elaborate, hill-sized tomb took hundreds of thousands of laborers decades to construct. Today it is an archaeological work in progress; the tomb itself never has been excavated. We do know that it is cardinally aligned. Moreover, we have a hint as to what we might find inside:

In 1974, local farmers were digging a new well. To their surprise, they struck an earthen head. It turns out that here, near the tomb of Qin, is an entire buried Chinese army in replica—thousands of soldiers, their weapons, vehicles, and horses—life-size. They guard Qin's tomb eternally, facing due east, the direction from which Qin's enemies came. Most of the figures are made of terracotta. The so-called terracotta warriors are one of the great archeological finds of our time.

Human ideas are often long-lived, even longer than the societies that created them. The symbolic seat of government in modern China, the famous Forbidden City, also is aligned cardinally.

RELATIVE MOTION

In modern times we know—or, at least, have read—that the motion of the stars about a north-south axis is an illusion. It is caused by the fact that we stand on a rotating, spherical earth. A star rises in the east because we are traveling (up to a thousand kilometers per hour) toward the east. (The earth rotates counterclockwise as viewed from the direction we call north.)

For a long time, people believed that the earth stood still and that all sky motion was relative to that static earth. Though incorrect, this is an entirely reasonable assumption. The rotation of the earth is counterintuitive. Even today, Neil Young can sing,

> This ol' world keeps spinnin' round
> It's a wonder tall trees ain't laying down

Relative motion is a funny thing. Have you ever driven a car to a stop between two tall trucks? When the light turns green, there is that odd moment when—just for an instant—it is unclear whether it is the trucks moving forward first or you moving backward.

An exercise: If you turn clockwise in a complete circle while observing your horizon, you will see familiar trees, hills, schools, and cell-phone towers. Each will enter the periphery of your vision, and exit it, in turn. But if you were not intentionally moving your body, you could imagine seeing exactly the same thing when, however improbably, everything on the horizon began to turn around you, counterclockwise, in unison. (There used to be an attraction at Disneyland based on exactly this illusion.)

Only a few hundred years ago, the majority of people still assumed that the earth was motionless, and that the objects in the sky circled it. It is a pretty straightforward conclusion from our perspective. It just happens to be wrong. Still, our language continues to acknowledge the misconception by including sky words like "rising" and "setting," implying that the earth is still and that the sky moves. There is no such set of simple English words for the reality of the situation. Instead of the word "rose," try to insert a single word into the following sentence that does *not* imply a stationary earth: "Just then, a bright star ____ above the horizon." There is none.

We are stuck with the language that we have. One way around this conundrum of relativity is to use a word meaning "apparent"—as in, for example, "Just then, a bright star appeared to rise above the horizon." (I already have begun to do so in this text.) However, this becomes cumbersome after a while. I will insert "apparent" occasionally, just to remind us. Most of the time, though, I will use relative words like "rise" and "set" without implicit literalism.

ZENITH AND CELESTIAL MERIDIAN

Vocabulary provides the convenience of compact sentences, which in turn can aid understanding. For instance, a circle of constant altitude on the sky is called an almucantar. "Almucantar" is a lot shorter than "circle of constant altitude on the sky." As altitude increases, the almucantar becomes smaller and smaller until it eventually becomes a point.

This point, directly overhead, is called your zenith. (The television or computer manufacturer of the same name must consider its product to be of the highest quality!) I will no longer take up space with "the point directly overhead," but will use "zenith" exclusively.

The zenith is by definition at altitude ninety degrees. Notice that it is the same angular distance from every horizon. The zenith is the only point in the sky with an altitude, but no azimuth. The zenith is ninety degrees from the horizon: one-quarter of a whole circle drawn all the way around you, from horizon to horizon to horizon.

Opposite the zenith, 180 degrees away, is your nadir—the point directly below you. As you might guess, this word has less utility in astron-

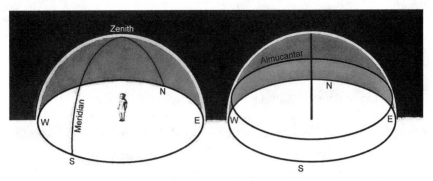

FIGURE 2.6. *Left,* The celestial meridian is a vertical circle with a particular azimuth. The word "meridian" (without the adjective) also is used sometimes to refer to a circle of constant longitude on the earth. I will try to avoid this ambiguity. *Right,* An arbitrary almucantar. An almucantar always is parallel to the horizon. Like the horizon, each point on an almucantar is equidistant in angle from the zenith.

omy than "zenith," but not zero. Both "zenith" and "nadir" are gifts of the Arabic language.

Notice I assume that the horizon delimits exactly half of all we can see beyond the earth. Would not a mountaintop astronomer see more than 180 degrees from horizon to horizon? This certainly would be true if we imagine the mountain elevation to be a significant fraction of the radius of the earth. In reality, the difference between Mount Everest and sea level is miniscule compared to the size of the earth. The difference in view from those two locations is negligible most of the time.

Astronomers do favor mountaintop observing, but not because they can see more sky from such a vantage point. (They cannot.) Nor is it because the mountaintop brings one closer to the stars. Stars are so far away that this difference in proximity is negligible, too. Astronomers do so mostly for another reason: On a mountain top, there is less atmosphere between them and a star than there is at sea level. See my box labeled "Extinction."

Back to our perusal of changes in the night sky as time goes by: Stars always reach their highest point (greatest altitude) halfway through their nightly partial circle across our sky, when they are on an imaginary north-south arc called the celestial meridian. The celestial meridian is a vertical circle that divides the sky into eastern and western halves. Stars increase in altitude east of the celestial meridian; they decrease in altitude west of the celestial meridian.

The zenith, the one point as far from the horizon as can be, helps us better define the celestial meridian. The celestial meridian is an imaginary half circle across the sky. Yet there is any number of arcs one can draw, north to south, across the sky. (Think of the segments in an orange.)

Only one of those, a vertical circle, passes directly overhead. That is the celestial meridian.

When a star crosses the celestial meridian, during the course of one day, its direction of travel with respect to the horizon changes. Instead of always increasing in altitude, once it encounters the meridian, the star decreases in altitude. As a star crosses the celestial meridian, it is said to culminate. Notice that, by definition, a culminating star is either due south or due north.

AN EXAMPLE OF CULMINATION

Do people notice culminations in the sky? They most certainly do.

I promise to give my students a dollar bill if they have on their person a picture of a pyramid. That is an unlikely thing to carry around, of course—unless you remember: There is a pyramid on the dollar bill itself.

The Egyptian pyramids hold a special fascination for people. At one time or another there have been those who have tried to see in them vast stores of lost knowledge, or all known laws of physics. There is a tiny piece of this modern folklore that rings true:

The Great Pyramid of Khufu (circa 2600 BCE), called "Cheops" in Greek, is the only one of the Seven Great Wonders of the Ancient World still in existence. Not only is it the earliest of the Old Kingdom pyramids at Giza, but it is also the biggest, occupying fifty-three thousand square meters of ground. Each side of the square base measures 231 meters, and the Great Pyramid was originally 146.6 meters high. (It used to be smooth and topped by a shiny golden cap.) It was the best built of the Giza pyramids, too: About 2.3 million huge stone blocks, each averaging 2.5 tons, were used in its construction. In comparison, all other pyramids are knock offs.

Not surprisingly, Khufu's is the pyramid that has been surveyed most thoroughly. The Great Pyramid is an early example of cardinal alignment: Its sides are exactly north-south and east-west. By "exactly" I mean within less than one tenth of a degree.

What about the pyramid's interior? There are multiple rooms within, but while the rest of the pyramid is limestone, one room is made out of granite. The nearest source of this stone is some eight hundred kilometers up the Nile. The room is called the King's Chamber and is thought to have at one time been the pharaoh's tomb. (There is still a lid-less granite-lined sarcophagus there; otherwise, any furnishings have been looted.)

From the King's Chamber, two narrow shafts slant upward to the exterior. One is at a thirty-one-degree angle to the horizontal, the other

Stars in
the Church

The concept of the celestial meridian may seem arcane today. If so, it may surprise you to know that it was honored in many great Catholic buildings of the Renaissance period and later. Specifically, the celestial meridian as it would be projected onto the ground was physically recreated for, aligned upon, and set into, the church floor. Historian John Heilbron has visited these meridian churches all over Europe. A few examples are the Basilica of San Petronio in Bologna, the Basilica of Santa Maria del Fiore, and the Duomo di Milano (each a cathedral). The celestial meridian appears in the Belvedere a Torre dei Venti (Tower of the Winds) within the Vatican. Some architectural meridians even had named stars symbolically laid along the meridian line, at their relative positions of culmination, north-south. The meridian in the Église Saint-Sulpice, Paris, is an (embellished) plot element within the pages of Dan Brown's 2003 novel, *The Da Vinci Code.*

Remember that the cathedral was once not merely a seat of ecclesiastical authority, but was also the center of community life. Just as the church served many purposes, so did the cathedral meridian lines: They were used calendrically to establish feast dates. They were used civilly to fix the hour. And they were used as scientific instruments by astronomers studying the orientation of the earth and sky.

is at a forty-four-degree angle. Because the King's Chamber is not directly under the apex of the pyramid, the two shafts end up emerging from the pyramid at about the same height.

The two passages originally were called air shafts. (After awhile, the mummy inside did not smell too good?) But that is farfetched: The pair is too small to be useful for ventilation, and nobody with a working nose was supposed to be in the tomb, anyway.

So maybe the "air shafts" were really passageways to the heavens. If so, where was the pharaoh heading?

Much later the New Kingdom Egyptians marked time through the night by certain stars, or groups of stars, culminating in succession. The description of one such group is unmistakable: a giant, standing man. Even today we recognize this as our constellation of the hunter, Orion. The unique thing about this constellation is the set of three evenly spaced stars in a line (which make up the belt of Orion) in the middle of the rectangle of bright stars forming Orion's shoulders and legs. (The belt of Orion is a familiar asterism, a distinctive star pattern that is not considered a constellation by itself.)

A central tenet of Egyptian religion in the later days (which we know more about), which was probably important in Khufu's time, too, was the idea that the soul of the deceased monarch would rise to the sky—not an unfamiliar idea—to meet up with the god Osiris for judgment. There are even references to the three stars that make up his belt. Osiris was Orion.

Astronomer Virginia Trimble and Egyptologist Alexandre M. Badawy calculated where these three stars were during pyramid-building times. At the latitude of Giza, where the Pyramid of Khufu is located, their altitude as they intersected the celestial meridian due south was forty-four degrees—the angle of the south shaft. The south pyramid shaft pointed, at the time of construction, toward the belt stars as they culminate once per day.

Maybe the Old Kingdom Egyptians meant some other stars, stars that appear over the shaft at another time (those east or west of Orion)? As we look around in all directions, there simply was not anything else, nearly as bright and distinctive, that passed over the shaft at the time of construction. The belt stars probably were the target.

Khufu is one of a set of three pyramids. The importance of Osiris in Egyptian religion has prompted some to speculate that the three pyramids themselves are a depiction of the belt stars. However, this is a weak argument: *Any* three artifacts lined up form a belt.

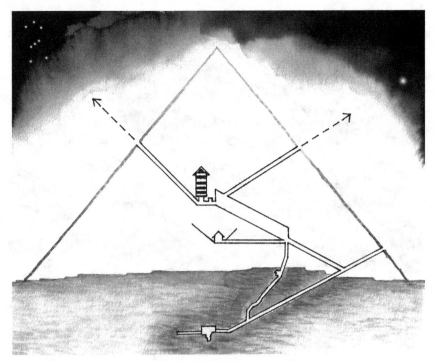

FIGURE 2.7. Cross-section of the Great Pyramid. The Egyptian polestar was Thuban.

CIRCUMPOLAR STARS

How familiar you are with the stars and constellations is likely a function of how often you see them in the sky. For most of us, some stars are more likely to be above the horizon at any given time than are others.

Still in the Northern Hemisphere's midlatitudes, we continue our tour of interestingly placed stars: A yet-more-northward-rising star will arc across our northern sky, not the southern. Close to the point about which the stars seem to turn, it may never set at all. Rather, it turns 360 degrees counterclockwise, a complete circle in one day, never dipping below the horizon. Such a star is easy to find; it is always in the sky. Any celestial object that behaves this way is said to be circumpolar from our location. The most famous circumpolar star is the North Star, which travels in such a small daily circle that it may not seem to move much at all. (Daily motion in the sky is sometimes called diurnal motion.)

The farther a star is from the pivot point, the longer the arc it will travel in one unit of time (a minute, an hour). Circumpolar stars may travel smaller angular distances in the sky than rising and setting stars do, but we can, in principle, see them move in complete circles during

FIGURE 2.8. Facing north in the Sonoran Desert. We see the trail of a bright star, through lower culmination and over the top of the cactus. Meanwhile, the North Star (brightest in the photograph) exhibits a tiny arc through upper culmination. This exposure was made over eight hours. (Because the North Star's light is spread over few pixels, each pixel absorbs more light than do each of the pixels illuminated by the other stars in the picture. The result is that the North Star looks brighter in the photograph than it does to the eye.) Photo by Joe Orman.

the course of one day—not just an arc. A circumpolar star culminates twice a day (upper culmination and lower culmination).

Remember the Great Pyramid of Khufu? After the pharaoh met up with Osiris in the sky (presuming he was judged worthy), where would he head next? I will give you a hint: The later Egyptians spoke of stars called the "imperishable ones." What kind of stars are "imperishable"? Circumpolar stars. They do not set/"die." Now, which one?

I already have written about the south-facing pyramid shaft pointing toward the stars of Orion's Belt. But there is still that north-facing shaft . . . Trimble and Badawy found that, at Giza, the north-facing pyramid shaft pointed right to the North Star—the ultimate circumpolar star. (See figure 2.7.) Khufu was on his way to join the Imperishable Ones.

Was the Great Pyramid an observatory? No. The shaft angles I have presented are averages; the shafts are not perfectly straight. In fact, there is no clear line of sight through them at all. They are symbolic. Nevertheless, the incredible effort necessary to construct something as huge and long-lasting as the pyramids nearly five thousand years ago demonstrates the incredible power of the sky in the Egyptian religion and picture of the afterlife.

Even today, the Great Pyramid is an engineering marvel. People debate how it might have been constructed using technology of the day. They argue whether the huge labor force necessary was paid or coerced. And it was all built around a circumpolar star.

Constellations of stars often are easier to identify than single stars. This is especially true if the constellation pattern is prominent, memorable, and made up of bright stars. If the constellation is circumpolar and lies far north, it may be just as useful an indicator of approximate north as the North Star itself.

All these conditions now combine in that part of the constellation Ursa Major known as the Big Dipper. A dramatic example of the Dipper's practical use can be found encoded in the folksong, "Follow the Drinking Gourd," sung by enslaved people in the antebellum United States. The song is a secret set of instructions for escaping slaves, directing them to freedom in the North. An explicit reference to our North Star (proper name: Polaris), or even Ursa Major, would be too obvious. It was best to maintain the fiction (in the presence of those who supported and maintained slavery) that African Americans did not know practical astronomy. So the drinking gourd was substituted into the words of the song. By following these stars north, and other directions hidden in the lyrics, a freedom-seeking slave could follow the stars and landmarks to connect up with the so-called underground railroad—and a life in non-slave-holding states. The story of the Drinking Gourd now appears often in children's literature. Christine Wilder has published a bibliography in *Astronomy Education: Current Developments, Future Coordination*, edited by John A. Percy.[4]

The two outer "bowl" stars of Ursa Major are called the Pointers. They are bright stars, and happen to point in the direction of Polaris (itself part of the Little Dipper, the constellation Ursa Minor). As they turn 360 degrees around true north, this asterism can be used to tell the time of night. Any device that indicates the hour by the orientation of circumpolar stars is called a nocturnal.

3

A Globe of Stars

By day you cannot see the sky
For it is up so very high.
You look and look, but it's so blue
That you can never see right through.

But when night comes it is quite plain,
And all the stars are there again.
They seem just like old friends to me,
I've known them all my life you see.

There is the dipper first, and there
Is Cassiopeia in her chair,
Orion's belt, the Milky Way,
And lots I know but cannot say.

One group looks like a swarm of bees,
Papa says they're the Pleiades;
But I think they must be the toy
Of some nice little angel boy.

Perhaps his jackstones which to-day
He has forgot to put away,
And left them lying on the sky
Where he will find them bye and bye.

I wish he'd come and play with me.
We'd have such fun, for it would be
A most unusual thing for boys
To feel that they had stars for toys!

AMY LOWELL, "The Pleiades,"
in *A Dome of Many-Coloured Glass*, 1912

So far, I have used a vocabulary that does a good job of describing what takes place every night in our sky: the horizon-zenith coordinate system. This vocabulary is made up of words such as "horizon," "zenith," "altitude," "azimuth," and "meridian." It is based on an imaginary bowl of stars rolling overhead. This reference frame is personal. By that I mean that it is centered on the observer—you. Your horizon is slightly different than my horizon, because we cannot stand on exactly the same place at the same time. Your zenith is slightly different than that of the person standing next to you. What is more, you carry your personal sky bowl around with you wherever you go. As you walk about, your zenith moves with you, forever hovering above your head. You can walk toward your horizon, but your horizon will continually stretch out of reach ahead of you. You will never get there.

Perhaps a better (and more modern) analogy of the sky, for the traveler, is an umbrella. As you carry an umbrella, its rim (horizon) always surrounds you, while its point (zenith) remains fixed atop your head.

As useful as the sky-bowl concept is, it is not universal. It can become confusing when two people at very different locations try to talk about the sky, because the definitions they use are not the same. An absolute way of talking about the sky is called for.

THE CELESTIAL SPHERE

In addition to the horizon-zenith system, there is another, parallel vocabulary set for describing what is seen taking place in the sky—one that is the same for everybody. It is called the celestial sphere. The celestial sphere takes into account the fact that the stars do not cease to exist when they have left our personal sky bowl. They are still there, merely below our horizon. For their apparent motion to be consistent, we must imagine a continuation of the bowl that we cannot see. It is logical to imagine an entire sphere of stars, not just half a sphere, that surrounds us. We can see only half of it at any given time.

Imagine that we live inside a giant beach ball. That is the celestial sphere.

Like our sky bowl, the celestial sphere is admittedly not a real thing. There is no big, finite shell of stars engulfing the earth—space seems to extend infinitely in all directions. Still, as we look up into space, it looks as if there is such a shell of stars surrounding us. The celestial sphere, too, is a handy concept for describing the overall motion of the sky.

Just like the horizon-zenith reference, the celestial sphere is "us centric." Only this time, "us" is everybody on the earth. As the celestial sphere is infinite in extent, the earth is infinitely small in comparison. It is as if the whole earth were reduced to a mere point at the center of the sphere. (This is why we see exactly 50 percent of the celestial sphere all the time, even though we are not at the center of the earth.) Because we are at the middle, the biggest circle that can be drawn on the celestial sphere is one centered on us—a great circle. (Our horizon is an example of a great circle, too.)

Each day, the celestial sphere appears to rotate once completely. The direction of progress is such that most stars appear to rise in the east and set in the west, just like the sun. (Circumpolar stars are an exception to this rule.)

All this is no coincidence. The sky appears to move in this fashion because, in reality, we on the earth are rotating under the celestial sphere—eastward. Spinning on a playground merry-go-round, you can make the landscape that encircles you enter your field of vision, rotate through it, and then "set." The result is the same: apparent circular motion around you while, in fact, only you are doing the moving. As you can tell by now, almost all movement in the sky, discernible over a night, can be explained by the simple fact of a rotating earth.

Those stars that rise and set southeast-southwest? Those stars are situated on the south celestial hemisphere. ("Hemi" just means half; this is the southern half of the celestial sphere.) The stars rising and setting northeast-northwest belong to the north celestial hemisphere. An east-west moving star sits on the celestial equator, which separates the two hemispheres.

The celestial equator is another of those imaginary circles or half circles we paint on the sky; this one happens to be at right angles to the celestial meridian. Living in the earth's Northern Hemisphere, the celestial equator is always in my southern sky. My northern sky is taken up completely by stars of the north celestial hemisphere.

The points about which everything seems to rotate are called the north celestial pole (always above the horizon for us northerners) and the south celestial pole (always below it). Just as was the case with the celestial equator, a celestial pole is an imaginary construct. There is nothing physically there.

Those familiar with maps will recognize that the geography of the celestial sphere simply borrows earthly geographical terms and adds the adjective "celestial." This is appropriate, as the celestial poles are just the extension of earth's spin axis, reaching infinitely into the sky. Similarly, the celestial equator is the projection of earth's equator radially outward, away from the earth.

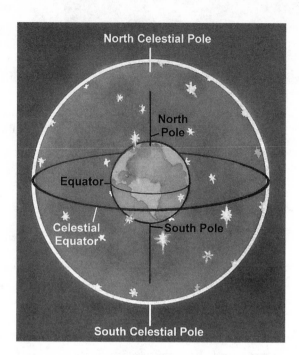

FIGURE 3.1. The celestial sphere depicted as a globe surrounding the earth. Instead of places, the celestial hemispheres, poles, and equator are better thought of as directions to which one can point from the earth.

Because stars are "fixed" on the celestial sphere, how far north or south a star is on the sphere will not change (for a long, long time). Stars closer and closer to the celestial pole will appear to move in smaller and smaller circles. If we can see a star, even at the lowest culmination on its circle, it is one of those circum*polar* stars. Now you know from where that term comes.

One real star, the so-called North Star or polestar, is so close to the north celestial pole that the rotation of the earth barely affects its position in the sky all night. It is the one star that does not perceptibly change its altitude and azimuth over time. The fact that it always is spotted in one particular place in our sky (e.g., over the garage, or between the oak trees in our yard) makes it easier to find.

For people living at midlatitudes, the celestial pole (north or south) is about half way up the sky. Of course, as every good Boy Scout and Girl Scout knows, the North Star is useful in finding geographic north.

There is no such well-placed star near the south celestial pole; that would be too much good fortune, I suppose. (But then, by geographical luck again, most humans live in the earth's Northern Hemisphere, not the Southern.) There are ways around this: Brazilian Indians found that, when the two "arm" stars of the constellation Crux (the Southern Cross) are on the same vertical circle and the brightest of the stars is below its companion, they are due south.[1]

FIGURE 3.2. The Magellanic Clouds. These features of the south celestial hemisphere appear as detached fragments of the Milky Way. They are unfortunately named, as they have nothing to do with the weather! They are vast, distant collections of stars. Photo by Akira Fujii/David Malin Images.

Just to remind ourselves that the apparent motion of the stars reflects the relative motion of the earth, not the stars, imagine staring at an overhead lightbulb. As you rotate your body in a circle, the lightbulb remains at the center of your view. But everything else, closer to the edge of your vision, appears to turn around the bulb like the hands on a clock. The bulb is the north celestial pole, or its nearby stand-in, the North Star.

Though they hold different stars and constellations, the north and south celestial hemispheres are largely the same as far as the unaided eye can see. One difference has to do with the Milky Way: Our solar system is not at the center of the disk-shaped galaxy. We are closer to one "edge" of the galaxy than to another. Gas and dust in the galactic plane obscure our view of the most distant stars past the galactic center. Still, the Milky Way is a bit brighter in the direction of the center. This center happens to lie in our south celestial hemisphere.

There are two satellite galaxies to our own called the Magellanic Clouds. They lie outside the plane of the galaxy. These are close and luminous enough that they can be seen with the naked eye. The Large and Small Magellanic Clouds are located in the south celestial hemisphere, so many northern dwellers are unfamiliar with these fuzzy patches in the sky.

(The Magellanic Clouds are so named because Ferdinand Magellan spotted them on his globe-circling voyage. However, the idea that he discovered them is absurd. They were well known to indigenous peoples of the earth's Southern Hemisphere long, long before Magellan ventured into their latitudes.)

CHANGES IN LATITUDE: NORTH

The Alaska state flag features part of the constellation Ursa Major (the Big Dipper). While this constellation is high and circumpolar in the Alaskan sky, it is invisible in Argentina. The Southern Cross adorns the national flag of New Zealand (among others). However, if you never leave the United States, this pattern likely will be meaningless to you.

The purpose of looking at selected imaginary—or not so imaginary— stars in the last chapter was to examine how and where stars at specific places on the sky bowl (or, if you prefer, celestial sphere) would appear from a specific spot on the earth through the cycle of the night. Now, let us study what happens when *we* do the moving. (Remember, we are starting in the midlatitudes of the earth's Northern Hemisphere.)

Here we go: The more northward we travel, the higher in the sky are the stars of the north celestial hemisphere. These north celestial hemisphere stars are spending more and more time above your horizon, too. Many are becoming circumpolar.

The North Star—the one star you trust to maintain its position in your home sky—now appears to be moving, too. It is increasing in altitude. Ninety degrees away, the celestial equator is decreasing in altitude.

On the other hand, less and less of the objects in the sky belong to the south celestial hemisphere. Those that you still can glimpse never get very high above the horizon. In fact, there are some stars, which you are used to observing back home, that here do not rise high enough to appear above your horizon at all. You will not see these southern stars until you return to your starting place.

We continue further northward. While you are getting a better and better look at the north celestial hemisphere, over your shoulder the south celestial hemisphere is disappearing from view.

Eventually we reach a place where the altitude of the north celestial pole is ninety degrees. The altitude of the celestial equator is zero—it lies along your horizon. All the stars you see are stars of the north celestial hemisphere. They are all circumpolar. The stars do not change in altitude as they dance parallel to your horizon, around an imaginary maypole extending to your zenith. The carousel-like appearance of the sky is quite different from that at midlatitudes. We have reached the earth's North Pole.

Why is nothing rising nor setting anymore? It is because, once you pause in your journey at the North Pole, *you* are no longer moving. The axis of the earth projects from your feet up infinitely through the zenith above your head. (Not as painful as it sounds.) You are at the pivot point. The rotation of the earth spins you, but no longer moves you from place to place. It is as if you stood in the middle of a scratch artist's turntable.

CHANGES IN LATITUDE: SOUTH

It is cold at the North Pole! Let us now travel south. (Can you think of any other direction you could choose, from the North Pole?)

Traveling south, everything happens in reverse. The celestial equator returns to the southern sky. The north celestial pole slips down from the zenith; it is now unambiguously north. Familiar constellations that were anticircumpolar in the far north reappear to say hello again. Some stars rise and set again. If we stepped off the North Pole just right, we will return to exactly where we started. Otherwise, we may end up at some strange longitude on the earth, but at the same latitude. It does not matter for our purposes: The appearance of the sky would be the same.

We do not have to stop here, however. As we travel further southward, the north celestial pole dips lower, and the celestial equator rises higher. Some unfamiliar stars appear in the south: These were forever anticircumpolar farther north; they never rose above our horizon. We are getting a better view of the south celestial hemisphere. Behind us, fewer stars are circumpolar. Eventually we reach latitude zero.

An observer on the equator has the advantage of seeing the entire celestial sphere as it rolls over her. The south celestial hemisphere takes up the southern half of the sky, and the north celestial hemisphere takes up the northern half. The celestial equator runs through the zenith; however, the celestial poles are now at the horizon. Nothing is circumpolar. We find ourselves in an enormous hamster ball, rolling east to west. Again, this is a different effect than what we watched each night at home. Eventually, we can see all the stars of the celestial sphere standing at the equator—though, of course, not all at once.

Journeying even farther south results in a mirror image of what we saw as we headed north: Just juxtapose the words "north" and "south." Stars in the north disappear, while southern stars become circumpolar—about the south celestial pole. (The north celestial pole is below our horizon from now on.) It certainly looks peculiar to our northern eyes to see most of the sky taken up by the south celestial hemisphere and the celestial equator in our northern sky. The clockwise motion of stars about the celestial pole seems odd, too.

Even familiar star patterns cause a double take far to the south: I remember one nighttime airplane flight to the Southern Hemisphere. The familiar constellation of Orion hovered out my cabin window. However, as we flew, this cosmic gentleman did a summersault and stood on "his" head. Wait a minute . . . Orion did not *do* anything; it was I who flipped upside down.

I leave the appearance of the sky from the earth's South Pole to the imagination of the reader.

NAVIGATION

Did you notice something in our travels? When we were at the North Pole, the altitude of the north celestial pole was ninety degrees. (It was at our zenith.) When we were at the equator, the altitude of the north celestial pole was zero degrees. (It was on our horizon.) The altitude of the celestial pole *always equals our latitude*. Through the good graces of some neat spherical trigonometry that need not concern us here, the smallest angle between the celestial pole and horizon is always equal to your latitude on the earth. For instance, if you live in Minneapolis, Minnesota, your latitude is coincidentally 45° N, and the north celestial pole is at an angle forty-five degrees above your horizon.

How handy for navigation! While measuring the altitude of the celestial pole (or its nearby surrogate, the North Star) might be difficult on a heaving ship's deck, the principle is, at least, straightforward. As long as you know what hemisphere of the earth you are in—and if you do not know even this, you are more than lost—the measurement of that single angle will tell you where you are. Knowing which celestial pole you are looking at tells you whether the latitude measured is north or south; if you really were so adrift as to not know upon which half of the earth you are sailing, the direction of circumpolar stars, counterclockwise in the north or clockwise in the south, would tell you.

The one number you need is the altitude measurement. No calculations are necessary. That is your latitude.

Determining one's longitude on the earth is a more difficult matter. And it is likely that a survey of ships foundering due to bad navigation will yield more vessels running aground east or west than north or south.

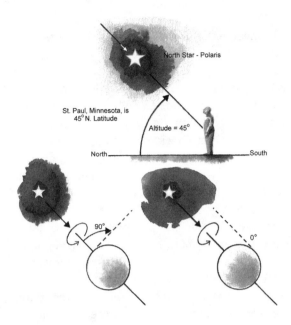

FIGURE 3.3. *Top,* Calculating the latitude of Saint Paul, Minnesota. *Bottom,* The celestial pole is at your zenith and horizon at the North Pole and equator, respectively.

A simple instrument used aboard old-time ships for measuring altitude was the quadrant. It consisted of a single plate in the shape of a quarter-circle disk marked off with ninety-degree increments along its partial circumference. (Think of one out of four evenly cut piece of pie.) A plumb bob hung from the right-angle corner of the plate. While the level horizon might be hard to establish on a pitching vessel, the bob could be relied upon to point toward the nadir (under gravity). So altitude was read from the plate where the plumb line crossed it.

Note: Remember, it is to be understood that everything I have said so far about north and south will be reversed for persons residing in the midlatitudes of the Southern Hemisphere. While this text is written primarily for persons who live in the Northern Hemisphere, a Southern Hemispheric reader could, in most cases, go through it, substituting the appropriate antonyms for these words.

STAR SEASONS

On any day, part of the celestial sphere is invisible. This is not just because it is below the horizon some of the time. It is because it is daytime some of the time, and the beaming sun turns the dark sky blue, thereby overwhelming all but the brightest astronomical bodies. A celestial object might be high in the sky, but that fact may not do us any good if it is noon. The myth that you can see stars better in the daytime by looking through a tube or (less conveniently) from the bottom of a well, is—well—a myth.

Now the ancients knew that the stars (those on one half of the celestial sphere) are actually "out" in the daytime. They understood that we simply cannot see them because they are blotted out by the light of the sun. So, on any given day, we can see roughly only half of the stars that rise above our horizon: The other half has been "covered up" by blue sky.

However, the sun itself appears to move across the sky with respect to the stars. It does so slowly, a little less than one degree per day, out of 360 degrees all the way around a great circular path. The result is that the sun reaches the same point on the celestial sphere once in 365 and a quarter days: the year. The reason for this is not literally that the sun is moving, but that the earth is revolving about the sun. Once again, relative motion yields an incorrect impression. And once again we resort to a useful fiction: the sun's motion around the celestial sphere.

As we earth inhabitants travel along with our planet, different parts of the celestial sphere are blocked from our view (by the sun when it is above our horizon) and different parts are exposed (when the sun is below our horizon). So superimposed on the effect of the whole sky seeming to turn once every (almost) twenty-four hours, due to the earth's axial rotation, is the fact that the sky changes nightly due to the earth's revolution about the sun.

Be careful of these two words, "rotation" and "revolution." In common speech they often are used interchangeably. Yet they mean different things. Rotation refers to something spinning on an axis. The spinning object may stay in one place. Revolution refers to traveling around another object, in a closed path. The path is a circle or some other continuous curve. In the case of a planet and the sun, this path is elliptical and called an orbit. The earth both rotates and revolves in the same angular direction.

The combined effect of the earth's rotation and revolution is that, if you go outside at the same clock time every night, you see that each star sets (approximately) four minutes earlier than it did the night before. (Our clocks are based on the sun.) Four minutes may not seem like a lot. Indeed, tonight's 10:00 p.m. sky will look a lot like last night's 10:00 p.m. sky—and tomorrow's. But four minutes plus four minutes plus four minutes . . . adds up. Over 365 days it amounts to a 366th. It is as if there is an extra apparent rotation of the celestial sphere in addition to the diurnal ones we expect each year.

This would not happen if our clocks were set to coincide with the apparent motion of the stars only. In that case we could choose any bright

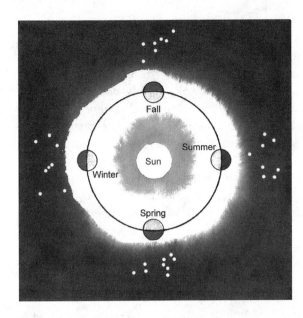

FIGURE 3.4. Some prominent constellations visible high in the night sky, for different times of year. The earth and sun are not to scale.

star—Rasalhague, let us say—and define the day (the rotation period of the earth) as the time from when Rasalhague culminates to when it does so again. What we would notice, though, is that the sun, as it appears to move eastward on the celestial sphere each day, is then *not* at the same place in the sky at the same time (perhaps the celestial meridian, at noon).

The sun is much more important to day-to-day affairs than are the stars. We want the sun to set in the evening, rise in the morning, and culminate at about noon—every day. This would not happen if we commonly measured time with respect to the stars. After a while, the sun would rise at midnight and set at noon. After that, it would get even more confusing.

So we measure our clock day with respect to the moving target that is the sun. Therefore, the sun is high in the sky at noon and nowhere to be found in the sky at midnight—every day of the year. (Some places there are exceptions to even this commonsense rule, which I will discuss later.)

Thus, it is the celestial sphere that seems to be placed differently in our sky each day, and not the sun. Imagine we are out looking at the night sky at 8:00 p.m., an arbitrary but convenient time. It is January. We see above our horizon a familiar set of stars and constellations that we in the north call the winter-evening sky. However, by April, the earth has traveled one-quarter of its way around the sun. There are stars that we used to see on January evenings that have been intruded upon by the sun. They are lost in daylight. On the other hand, there are now stars that we could not see in January (because of the sun) that are now visible in the April evening sky, absent the sun. We are presented with the stars and constellations of the spring-evening sky. In July, our night sky (the summer-evening sky) is the portion of the celestial sphere that was in the same direction as the sun last January—it was in the daytime sky— but now is in that part of the celestial sphere opposite the sun. It is our current nighttime sky. In other words, at 8:00 p.m. in July we see the opposite celestial hemisphere from that which we saw at 8:00 p.m. in January. Daytime and nighttime skies are flipped. The earth moves another quarter of the way around the sun. By October, the month in which I write these words, more stars have been "lost" (and "found") to the evening sky. The stars and constellations above our horizon are those we associate with the autumn-evening sky. With patience, over the course of the year we are able to see in our evening sky all the stars and constellations that ever rise above our horizon.

Of course, if you work third shift, you might be more interested in what the sky looks like in the early morning. We could just as well talk about the winter, spring, summer, or autumn morning skies. They, too, would be the same each year—something to which we can look forward. And if you observed the sky all year around, you still would not miss a thing.

AN EXAMPLE OF RISING AND SETTING USING THE CELESTIAL SPHERE MODEL

Thinking of the sky as the intersection of the horizon with a celestial sphere makes clear why where on the horizon a star rises and sets is dependent upon how close or far it is from the celestial poles. Only a star equidistant from the celestial poles (i.e., on the celestial equator) can ever rise due east or set due west. All the rest of the stars (except circumpolar stars, of course) must rise either northeast or southeast and set either northwest or southwest.

We now go to the New World, specifically the abandoned city of Teotihuacán.[2] When we think of ancient Mexican civilization, we usually think of the Aztecs. But Teotihuacán was not built by Aztecs, nor even their predecessors, the Toltecs. It was built by people who lived on the Mexican plateau before the Aztecs and Toltecs, two to three thousand years ago. We know them only by the name of their great city: Today they are called the Teotihuacános. (We do not know what they called themselves.) They were the first Mesoamerican urban center.

The Teotihuacános were the Roman Empire of North America. In fact, their city was bigger than Imperial Rome. Teotihuacán covered over thirty square kilometers. Some have estimated its peak population at 250,000. Though nobody in the Old World knew that it existed, it was once the third largest city on the planet.

And Teotihuacán is a planned city. An urban plan causes those living in the outlying regions to remember their connection to the city proper. Take, for instance, Glendale, Arizona, where I grew up. Glendale on the west and Scottsdale on the east continue the street-numbering system used in the bigger city between them. When I was young, it was always clear that these two suburbs were bedroom communities for the state capitol, Phoenix.

Everything in Teotihuacán is aligned to a central avenue, now called the Street of the Dead, and another so-called avenue at right angles to it, the two of which divide the city into four quarters. Near the city center are aligned pyramids (pyramids again) now called the Pyramid of the Sun and the Pyramid of the Moon. Teotihuacán is an aligned city: Its inhabitants even modified the course of the local river to flow in alignment with the city. Nobody was doing anything this elaborate in Europe up to this time.

The Pyramid of the Sun stands seventy-five meters high and is 220-meters wide at its base. This is comparable to what we found in Egypt: The Pyramid of the Sun is the same base size as the Great Pyramid of Khufu (but only half as high). Indeed, the Pyramid of the Sun is the third biggest pyramid on Earth.

Cardinal alignment is so common among planned cities that we notice its absence more than its presence. So the surprise at Teotihuacán is that the Street of the Dead (called this because the later Aztecs thought the pyramids were tombs) does *not* run north-south. It is skewed fifteen and a half degrees to the east of north.

What happened? Did the builders screw up? Unlikely. Archaeoastronomer Anthony Aveni has discovered what he believes to be the baseline used to build the city: It is tied to a pecked-cross design. One such petroglyph shows up south of the Sun Pyramid itself, another is placed on a hillside three kilometers away. Together they make a perfect ninety-degree-angle line to the Street of the Dead. Compromising this idea that the Teotihuacános used such a sophisticated surveying technique as laying out a baseline is that more than forty more pecked crosses subsequently have been found that do not necessarily align to anything.

Still, everything is tilted fifteen and a half degrees. This suggests something celestial. Knowing, by archaeological means, about when Teo-

FIGURE 3.5.   The Pleiades. Only the brighter stars can be seen without a telescope. Courtesy of NASA, ESA, and AURA/Caltech.

tihuacán was built, we can run the sky through the day and year, using a planetarium or computer program, and find to what Teotihuacán is aligned—if anything. It is a dangerous technique because it is apt to yield "false positives," but we will proceed cautiously.

On which axis do you suspect we will have the greatest chance of success? Because it is so close to north, the "almost north" axis of Teotihuacán circles little celestial real estate during the course of the day. (The circumference of a fifteen-and-a-half-degree-radius circle centered on the north celestial pole is small.) However, the "almost east-west" axis does cover a lot of territory on the celestial sphere. (Because it is fifteen and a half degrees plus ninety degrees in azimuth, this axis traces a circle that has nearly as large a circumference as the celestial equator, a great circle.) Even so, there is only one significant celestial object that Teotihuacán points to, yet it is totally unique.

In our northern winter-evening sky, near the constellation Orion, there is an asterism of six or seven stars closely spaced with similar brightnesses. They are the Seven Sisters or Pleiades. There is nothing else in the sky quite like it. Aveni found that, at Teotihuacán when the city was built (about 150 CE), the Pleiades set within one degree of the east-west line. (There is a pretty clear view of the horizon in that direction at

Clusters of Stars

The Pleiades (in the hindquarters of the bull constellation Taurus) is an example of a star cluster. An experienced eye under optimum conditions can see many more stars than the six or seven traditionally ascribed to the Pleiades. The telescope reveals still more, less luminous ones.

A cluster of stars is distinguished from a constellation because the stars in a cluster are physically close to each other in space. They are not a coincidental arrangement of stars, at different distances, near the same line of sight.

Other star clusters that you can spot with the naked eye include the stars making up the fuzzy part in the constellation Coma Berenices. The "V" in Taurus (the bull's head) is part of a cluster named the Hyades. The Praesepe, also called the Beehive Cluster, can be made out in the constellation Cancer.

Pleiades trivia: The Japanese name for the Pleiades is *Subaru*. We see this star cluster frequently on the street. It is stylized in the hood ornament used by the automobile manufacturer of the same name.

What about Accuracy?

When I discuss celestial alignments, I will not worry much about great accuracy with respect to that which might hypothetically be achieved with the human eye. That is not the point. To do so would be to place ourselves in a distinctly modern, scientific mode of thinking, one in which quantitative measurement is important. Such a worldview may be anachronistic to the people who produced such alignments. Rather, it is my goal to point out places where people, often long ago, clearly *acknowledged* the existence of potential alignments.

Why might an alignment be inaccurate? Foresights and backsights might be moved in azimuth through any number of natural and not-so-natural processes over the years and centuries. Precession, discussed in chapter 4, causes a systematic error with time (though one that can be accounted for). The difference between the true horizon and the astronomical horizon can affect alignment azimuths; the true horizon may actually *change* (trees grow, trees are cut down), thereby introducing apparent inaccuracy. Still other factors having to do with our atmosphere are inevitable. (See the box on extinction in chapter 6.)

Teotihuacán.) Notice that the azimuth of the Pleiades setting is determined simply by how far it is from the celestial equator (and how close it is to the celestial pole).

Why change the alignment of a city, from the more obvious cardinal, to stellar? While the Pleiades have no tangible effect on humans, obviously the sun does. And there was a strong connection between the Pleiades and the sun for the Teotihuacános. Of course, the Pleiades rise every day. Its special behavior at Teotihuacán was that it rose "heliacally," for the first and only time each year, *on the same day* that the sun reached the zenith. (Later I will discuss what the word "heliacal" means; for now, suffice it to say that it is an auspicious rising.) This is a coincidence of note. In the next chapter we will delve into the significance of the zenith for those, like the Teotihuacános, who live in the Tropics.

Other theories have been put forward for the Teotihuacán alignment. Some say the alignment is to the sun itself. In fairness, I should mention a nonastronomical hypothesis, too: The Pyramid of the Sun was discovered to have been built over a cave, and the axis of that cave is approximately that of the city axis. Nevertheless, it is the Pleiades orientation that I find most compelling.

4

Of Precession, Planispheres, and Patience

For who so list into the heavens looke,
And search the courses of the rowling spheares,
Shall find that from the point, where they first tooke
Their setting forth, in these few thousand yeares
They all are wandred much; that plaine appeares.
For that same golden fleecy Ram, which bore
Phrixus and *Helle* from their stepdames feares,
Hath now forgot, where he was plast of yore,
And shouldred hath the Bull which fayre *Europa* bore.

EDMUND SPENSER, *The Faerie Queene*, 1590

Up to this point, I have described the stars as if they were mounted on a celestial ceiling. It is a moving ceiling, but a permanent one. This is a fair analogy. Changes in stellar magnitude are subtle or nonexistent; changes in stellar color take place over time spans far, far greater than those with which humans normally contend. In this final chapter on stars, I will review the exceptional ones that *do* change in brightness; I will discuss a long-term change on the celestial sphere that affects the appearance of all stars in our sky; and then I will offer some practical advice on enjoying the fainter celestial objects, such as the stars.

CHANGING STARS

The stars, at first glance, appear to be constant lights in the sky. While many stars do vary in brightness, comparatively few vary so much that the unaided eye can detect the change.

A notable exception is Algol, in the constellation Perseus.[1] (It is also called β Persei.) Every few days, Algol dims noticeably for about ten hours.

This is not because the star varies intrinsically in brightness. Algol has an orbiting companion star that cannot be resolved (separated from Algol) by the naked eye. Algol is, in fact, a binary star. By coincidence, the orbital plane of this second, dimmer star intersects the earth. So once per revolution, the companion passes behind Algol and, instead of seeing the combined light of two stars, we see the light of only one. Their combined brightness dims. Also, once per revolution, the smaller companion star passes in front of Algol. It blocks out some of Algol's light. As the surface brightness of the companion is less than that of Algol, again the combined brightness of the pair dims. The second brightest star in the constellation Lyra (β Lyrae) is an eclipsing binary star, too.

With the advent of the telescope, astronomers noticed other stars that change in brightness. In retrospect, some of these variations were noticeable to the naked eye.

The star δ Cephei (in the circumpolar constellation Cepheus) is a more common kind of variable star. Here a single star pulsates, changing its size and luminosity. Delta Cephei dances to a rhythm, of brighter to darker to brighter, every five days. Another example is η Aquilae, whose rhythmic period lasts seven days. Mira (or o Ceti) reaches a maximum brightness that is usually visible to the unaided eye, about every eleven months. The brightest variable star is Betelgeuse (the brightest star in Orion, α Orionis). However, the variation of Betelgeuse is irregular in period and modest in amplitude.[2]

(Notice the use of Greek letters to designate stars that do not have proper names. Long ago, astronomers simply listed the star they thought to be the brightest in a constellation as alpha [α], the second brightest as beta [β], the third gamma [γ], etc.—though not always consistently. If you were ever in a fraternity or sorority, you know this to be Greek alphabetical order. "Cephei" is just the genitive case for the noun "Cepheus," "Aquilae" for "Aquila," "Ceti" for "Cetus," and so forth. The system stuck.)

Occasionally, a star normally too dim to see without a telescope will brighten markedly and unexpectedly. This sudden stellar eruption is called a nova. If the star becomes visible, it will look as if a new star has appeared on the celestial sphere. ("Nova" is Latin for "new.") Eventually the nova fades; the "new" star disappears.

Supernovas are rare, but spectacular. Novas may repeat; a supernova is a blazing, one-time phenomenon. It is an exploding star that may, over the course of a few hours, become brighter than all the other stars on the celestial sphere put together. There are reports of supernovas such that one could read by their light at night. Only seven nearby supernovas have been reliably recorded in all of human history: in the years 185, 393, 1006, 1054, 1181, 1572, and 1604. The astronomical telescope was invented in 1609. (Bummer.) For the first five cases, we are indebted to Chinese astronomers, who better documented unexpected astronomical events than did their counterparts elsewhere.[3] They gave us enough information that each of these supernovas, along with the pair from the sixteenth and seventeenth centuries, can now be identified with a supernova remnant (an exploding cloud of gas) visible through modern telescopes.

One other moderately bright, naked-eye supernova was visible in 1987: It was not dazzling because it was in the Large Magellanic Cloud and much farther away than the previously mentioned examples.

Is the 1054 supernova also recorded in Native American rock art? The case has been made—and widely publicized—for a famous rock painting in New Mexico: It shows what looks like a star next to the crescent moon. The supernova indeed would have appeared as a star next to the crescent moon on one night that summer (based on the Chinese chronology of the event). However, there are other bright "stars" in the sky. The planet Venus, for example, often can stand brilliantly next to the crescent moon; such a scene would be indistinguishable (at least on a rock painting) from the supernova-and-moon scene. Rock art is hard to date. How much more often has the Venus-moon configuration appeared in the sky during the centuries since available to paint the problematic pictograph? Other symbols painted in the vicinity make it reasonable to ask if this art has much to do with real events in the sky at all.

THE EARTH WOBBLES, BUT IT WON'T FALL DOWN

Notwithstanding a few counterexamples cited above, so far I have treated the celestial sphere as eternally unchanging. This is not true. Recall that the stars are not glued to the sphere; they are real objects moving through space. Yet stars are so far away that, even though they travel quite swiftly, it takes an incredibly long interval of time for them to appear to change positions in our sky. In fact, the time it takes is long, even when compared to the age of civilization. Few such proper motion stars would be identifiable by the naked eye, and then only by members of one generation of sky watchers comparing star charts with others of another generation.

Think of an aircraft in the sky: A distant plane appears to move most lazily, taking minutes to go from horizon to horizon. Yet if you have ever ridden on one, you know (especially from your experience at takeoff and landing) that jets go hundreds of kilometers an hour.

So I was not telling a falsehood when I wrote that the stars are "fixed" with respect to each other. This has remained true, insofar as the naked eye is concerned, over at least a hundred thousand years.

Nevertheless, the *entire* celestial sphere does appear to change its pivot point over a long interval of time. Right now, the north celestial pole points to a particular place in the sky that happens to be near the bright star Polaris (α Ursa Minoris). This has been true for a long, long time (all of our lifetimes and those of our parents and their parents, etc.). However, it has not always been the case, and eventually will not be the case anymore.

Besides its apparent rotation, there is another motion of the celestial sphere that affects the way the sky looks over very long periods of time. This is called precession. The axis of the earth actually wobbles, like a top. The north celestial pole (as well as the south celestial pole) marks out a circle in the sky. The reason why you probably have not heard of this motion, unlike the diurnal motion, is the time involved: It takes twenty-six thousand years to make one complete turn through the precessional circle. This interval is called the Platonic year (a misnomer because it is independent of the revolution period of the earth, the real year).

Think about watching the top. We easily can observe the wobble, and it usually is slow enough so that we can time how many wobbles occur in, say, one minute. At the same time, the top may be spinning so fast that its motion is a blur: You cannot even count the hundreds of rotations per minute the top makes. In this example, the ratio of the period of the top to the period of its wobble is hundreds to one. Yet think of how many diurnal rotations the earth makes in twenty-six thousand years. Obviously, it is an even much more extreme ratio.

While the day and year mark comparatively short periods of time relative to the human lifetime, the Platonic year does not. From year to year and, particularly, day to day, we cannot tell that the position of the pole is slowly moving. It is not a untruthful to tell our kids that Polaris marks the direction of north. This will be true for all practical purposes in their lifetimes and their children's and *their* children's and . . . but eventually, the north celestial pole will be moved noticeably away from Polaris (many hundreds of years from now), and some other nearer star may have to serve as the North Star. The point in the sky about which everything seems to turn will be nearer this new polestar than to Polaris. In twelve thousand years, for instance, the bright star Vega (α Lyrae) may serve to indicate true north. At the same time, the bright southern star Canopus (α Carinae) will become a reasonable "South Star." So subtle is precessional change in a human lifetime that it is a triumph

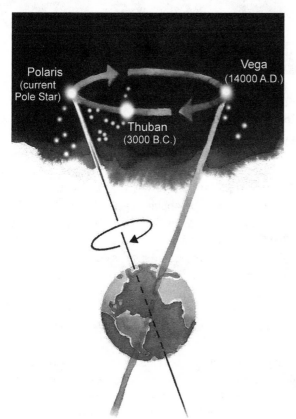

FIGURE 4.1. Precession. The direction of precession is opposite to that of the earth's rotation.

of naked-eye astronomy that precession was discovered, by a Greek astronomer named Hipparchus, circa 150 BC.

No, precession does not affect much in a practical sense—unless you are dealing with the lengths of time familiar to archaeologists. Remember when I wrote that the Khufu pyramid's "airshaft" pointed (past tense) toward the North Star? That star was not Polaris, it was Thuban—the brightest star in the modern northern constellation Draco (α Draconis). Thuban was the North Star in the time of the Old Kingdom Egyptians. In fact, it was about as close to the north celestial pole as Polaris is now.

An interesting thing about North Stars is that there does not have to be one. There will not be another comparably bright star near the north celestial pole for five thousand years. Then it will be Alderamin (α Cephei), and that star still will be a few degrees away from the north celestial pole at its closest. Indeed, there is no close-to-the-pole star (other than Thuban and Polaris) most of the time. The fact that there happened to be one in 2700 BC underscores why the ancient Egyptians might have been interested in the north celestial pole. The fact that there is today a North Star, so as to make the concept meaningful in our discussion, smacks of delightful coincidence.

How could one find north at night if there were no North Star? One way might be to pick out two circumpolar stars, one above the north celestial pole and one below it, and wait for the two stars to have the same azimuth. (A line connecting the two stars will be perpendicular to the horizon; a plumb bob will attest to this.) Mark this azimuth (by placing a foresight post on the horizon, for example). Then wait for the stars to apparently rotate 180 degrees. The first star now will be below the north celestial pole, and the second star will be above the north celestial pole. Once again, they will share an azimuth. Mark this azimuth (with another post or some such tool). True north will be at an azimuth half way between the two posts. Of course, if one of the two posted events occurs in the daytime, you will have to wait some months until it occurs at night. This is *a* method—it is hard to say that it is a *practical* one.

Let us return to precession. Notice that precession may theoretically help us date events. Other archaeological techniques yield the construction date of the Great Pyramid. However, what if we did not have these methods at our disposal? Instead, what if we only had the suspicion that the pyramid was aligned to a North Star? We could simply calculate when Thuban—the likely candidate—was in the right place to be in the direction of the pyramid "airshaft" (within a century or two). We could then hypothesize that this date was the date of construction (or, at least, design) for the pyramid. It would be best to leave this conclusion as a hypothesis, though.

Precession does not affect circumpolar stars only (and which stars will be circumpolar stars); it affects the entire celestial sphere. This includes the rising and setting azimuths, and the culminating altitudes, of stars that are not circumpolar. Another way to (very) tentatively date the Khufu pyramid would be to figure out when Orion's belt stars were in the direction of the south shaft—assuming that we already strongly suspected a connection between the pyramid and Orion.

Luckily, these Orion stars are far from the poles of precession: Stars near one of the poles of precession will not change their location much at all on the celestial sphere, even over lengths of time comparable to the Platonic year. Still, most stars are noticeably affected by precession.

What good fortune! If the precession period had been much longer than it is, it would be of no use at all to archaeologists who are interested only in time intervals comparable to the age of human civilization.

If the precession period was much shorter than it is, there might be an insurmountable ambiguity as to how many revolutions about the pole of precession had been made between the event we wish to try to date and the present.

For example, if the precession period was a mere thousand years, Thuban would align with the Khufu "airshaft" every thousand years. There would be several possible construction dates occurring during the long time span of human habitation in Egypt. In twenty-six thousand years, humans have gone from precivilization (a time from when we have no monuments to date) to modernity. For dating events (at least, theoretically) by the celestial alignments they incorporate, twenty-six millennia are just about right. (I say "theoretically" because I know of no date accepted by historians or archaeologists determined from precessional arguments alone.)

One more example: Precession affects where on the horizon the Pleiades set at Teotihuacán. By the tenth century CE,[4] the city axis no longer pointed toward the Pleiades due to precession. The civilization of the Teotihuacános was dying out by then, anyway. The Aztecs would call the ruins of Teotihuacán the birthplace of their gods. Anthony Aveni suspects that subsequent Mexican Plateau civilizations were so impressed with the empty city of Teotihuacán that they aligned their cities the same way. This was true, even though the significance of the orientation was lost to the newcomers. Some of these "copy" cities are to be found a hundred kilometers from Teotihuacán. Their locations and later construction dates (in the precession cycle) caused their alignments to deviate in azimuth from fifteen and one half degrees.

Precession eventually destroys coincidences between star phenomena and sun phenomena. For instance, if you mark the date to sow seeds by the rising of certain stars at dawn, there will come an era when you (more likely, your descendants) will plant at the wrong time of year. You could modify the rule—replace "dawn rise" with "dusk set," for example—and extend the tradition for many centuries. Still, the time will come when the calendar stars must be replaced, or your people will suffer the agricultural consequences. All such folk practices inevitably must die out.

It has been a long time coming! I can now summarize the five—and only five—things that determine the stars you will see in the sky. They are in the order of extent to which they most likely affect our view.

1. The rotation of the earth once per day
2. Your location on the earth (latitude)

3. The revolution of the earth about the sun once per year
4. Precession of the earth's axis
5. The rare appearance and disappearance of a star from and to naked-eye visibility

While all this apparent change in the sky may seem dizzying, remember that the apparent motions caused by the rotation, revolution, and wobble of the earth are readily predictable and always circular. Now that we understand the motions of the celestial sphere, we only have to add the motion (if any) of an object itself on the sphere. We will do so in chapter 5.

THE PLANISPHERE

Confused? Still having trouble with what's "up" and what's "down"? A device called a planispheric vovelle (or just planisphere, for short) may solve this problem. It is simply a star map mounted on a wheel with a window in it. By turning the wheel, dates on the chart can be matched up with times. To use the planisphere, dial up a date and hour. That is all there is to it: Unless you intend for the planisphere to become a family heirloom, you can once again ignore precession.

The planisphere's window now shows you the appearance of the sky at the chosen date and time. Its edge is your horizon; its center is your zenith. Hold the planisphere over your head and orient it. (North, south, east, and west should be marked on the planisphere.) You can match up the stars in the sky with the labeled constellations on the map.

You know that your sky is latitude dependent. Still, most of these devices are designed for anyone living at midlatitudes in the Northern Hemisphere. (They are good plus or minus five degrees—and maybe a little more—from the latitude for which they were established.) Southern Hemisphere planispheres are a little harder to find.

The planisphere is based on an ancient astronomical device, the planispheric astrolabe. The planispheric astrolabe may, in turn, date back to the Hellenic Greeks. The main difference between it and today's planisphere is that the moving part of the planispheric astrolabe was not mostly opaque. It was a complicated matrix of pointers that indicated the positions of stars, on the map backing, as they appeared at different times. You had to figure out where the horizon was on the planispheric astrolabe yourself. The astrolabe map looked like that which might be produced today by shining a light through half a transparent globe of the

FIGURE 4.2. Planispheres. Photo by Michael Hockey.

celestial sphere. The shadows of the stars painted on the globe would project onto a screen the way that they appeared on the astrolabe map.[5]

It was hard to pick out constellation patterns on the planispheric astrolabe. They would be "squished" far from the center of the map. Our planisphere uses a projection such that two stars (say) ten degrees apart, near the center pole, have the same physical separation on the map as do two other ten-degree-apart stars, nearer the celestial equator. Some constellations close to the horizon of the planisphere end up looking stretched out, but at least they are recognizable. In the next chapter, I will discuss more of the advantages and disadvantages of portraying a sphere on a flat plane.

Modern planispheres are inexpensive—paper ones more so than plastic. They sometimes have labels like Star Wheel or Star and Planet Finder. Look for them at science/nature specialty stores or hobby shops, in science or natural-history-museum gift shops, or in science-related gift catalogs. All of the above may have websites.

Some periodicals provide horizon-rimmed, planispheric maps of the sky, accurate for a given time, month, and latitude. (In a sense, they turn the wheel of the planisphere for you and freeze it in place.) These include *Astronomy* and *Sky & Telescope* magazines. Some astronomy texts and general trade books on astronomy include collections of planispheric star maps.

Before we leave the subject of stars and move on to the sun, moon, and planets, a few words about brightness: Most stars are not bright. A trick astronomers use to see stars at the limit of their sight is called averted vision. The retina of the human eye is most sensitive to fine detail in its center. I am sure this was evolutionarily useful: If early humans stared across the veldt and saw a charging lion, it would be best to be able to do so while the lion was still far away. While the center of the retina is designed for resolution, it is less sensitive to light than is the retina's periphery. By glancing slightly away from a star, you can use the more sensitive outer retina to make a dim object look brighter. Try it. It probably only will work for a moment—until your brain overrides your instruction to the eye not to look straight ahead in a normal fashion. But in that instant the faint star will "jump out" at you. It is as if somebody turned up the dimmer switch. This is almost startling. Averted vision works for other faint astronomical objects, as well.

Do not be too disappointed if you do not see star color. Individuals' color perception varies markedly. If you cannot see color in the stars, you are unlikely to see it elsewhere in the sky, either. The eye needs a certain amount of light to trigger color vision at all. (This is why, when you wake up in a dark room at night, everything looks gray.) Even the bright stars may not be enough to trigger your particular color vision—even if you are not diagnosable as color blind.

Be patient. Exactly when you observe is not as important as the length of time you spend doing so. If you have just come outdoors from a brightly illuminated building or vehicle, the iris of your eye has shrunk to a small size. Not much light is entering through your pupil. To discern faint stars, it is necessary for as much light as possible to strike the retina in the back of your eye, within the thirtieth of a second or so it takes the human brain to construct an image. When you first step into the darkness, it takes a while for your iris to open. In fact, up to twenty minutes are required for people to become fully adjusted to the dark. (Chemical changes must also take place in the eye in order to establish night vision.)

I think that not waiting for dark adaptation is the second most common cause of "sky disappointment." Plan on spending at least half an hour or so looking up in the dark to make sure that your eyes become fully light sensitive.

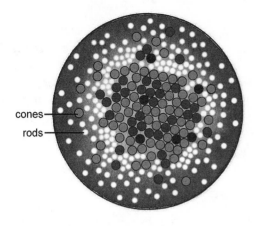

cones

rods

FIGURE 4.3. The inside of the human retina. Rods and cones are light-receptive cells. The cones are sensitive to color, but the rods are more sensitive to brightness than the cones.

The American Astronomical Society (AAS) lists other factors influencing dark adaptation. I quote:

Age: On the average, dark-adapted vision of the young exceeds that of elders.

Diet: Vitamin A is known to benefit vision in general and dark adaptation in particular; other beneficial items are quinine and carbon-sulfide compounds.

Physical Condition: Dark adaptation usually improves as general health improves.

Eye Condition: The better the general eye condition, the better the dark adaptation.

Heredity: An individual's genes may transmit the characteristic of good (or poor) dark adaptation.

Practice: Some evidence indicates that individuals who frequently change between high- and low-light levels in their work develop an ability to adapt more quickly than do others.

Preconditioning: . . . The higher the level of environmental illumination prior to entry into a low-level environment, and the longer the exposure to high-level illumination, the longer the time required to achieve dark adaptation . . . Rapidly pulsing light sources, such as fluorescent tubes which flash on and off 120 times per second, significantly prolong adaptation time to low-light levels as compared to steady sources, such as incandescent lamps of the same brightness.[6]

Stop smoking. Cigarettes impede dark adaptation.

One thing that the sky watcher cannot do is fight the effects of aging. Throughout life, the human pupil shrinks, though how much and when is specific to the individual. As a rule, though, a dark-adapted child can see fainter stars than can, say, a dark-adapted, tenured university professor. Compounding this effect is that the lens in the human eye becomes less transparent with age. It also becomes yellower, introducing a color bias. As for the cornea, it may dry with time and become irregular, thereby increasing deleterious light scatter. And all this assumes a healthy eye. Ultraviolet exposure speeds up eye aging and the onset of some eye diseases.[7] Wear your (UV-filtering) sunglasses.

WHERE TO OBSERVE THE SKY

Realistically, most people will observe the sky from near their homes. And as long as a given astronomical object is in your sky, there is a more important factor than latitude or longitude in selecting your observing site.

I cannot stress enough the importance of choosing a *dark* location from which to view faint astronomical objects. By "dark" I mean one far from artificial illumination of any kind. Those who were let down by such highly publicized temporary sky phenomena as Comet Hyakutake (1996) or Comet Hale-Bopp (1997) usually were those who did not make the effort to find such a site.

These days, finding the ideal viewing location is not as easy to do as it once was. Even though Halley's comet was not as favorably placed in our sky in 1986 as it was during its previous 1910 apparition, expectations ran high. After all, our grandparents told us what a sight it was and not to miss it. But our grandparents had an advantage over us that had nothing to do with the comet.

All over the world, skies were darker in 1910. Outdoor lighting in this age was in its infancy. We were much more a rural civilization. It was easier to find a dark place back then; it might even have been your own front yard.

Today, we are losing the dark of night. For most urban dwellers, even the brightest stars have all but disappeared. They have not gone anywhere—the fault lies in street lamps, spotlights, glowing fast-food signs, and so on. Astronomers call this problem light pollution.

Now, I have nothing against light. Light that lets us find our way at night is good. It keeps us from bumping into each other. I have no wish to return to the days when we were virtually prisoners in our own houses, once the sun had set.

Everything is fine as long as artificial illumination is directed where it is intended: down onto the ground. However, too much of this lighting is carelessly directed upward. This includes lights intentionally beamed up for advertising purposes and, more commonly, improperly aimed street and yard lights. This escaped light does not just disappear as it shines upward. It scatters in the earth's atmosphere and makes the sky glow, thereby obscuring faint astronomical objects.

Photographs of the earth at night, taken from space, show this problem clearly. North America and Europe particularly are ablaze with light. The location of every city and village can be pinpointed. The paths of interstate highways can be traced just by the glow of roadside rest stops.

This light is not doing anyone any good. It is totally wasted. Need I add that it takes energy to illuminate the stratosphere superfluously? Energy conservation begins above our heads. More pragmatically, all this light that never reaches the ground represents wasted money. In the case of public lighting, it is most often at the taxpayers' expense.

Do not get me wrong. I understand the need for security lighting to make our public spaces safe. Yet more light is not synonymous with better light. Efficient illumination involves putting enough light where it will be useful, and no light where it will not.

The irony of light pollution is that, unlike many other forms of pollution, the problem is so easy to cure. Often, that solution is as simple as placing a reflector over the lamp bulb to bounce light back down instead of beaming it up. Meanwhile, though, light pollution continues to rob us of a birthright no less legitimate than clean air and water: the right to enjoy a truly dark and beautiful night sky.

Let me address the subject of how to deal with light pollution. You can fight or flee. To fight light pollution in the big picture, you can encourage governments and businesses to install responsible illumination. More locally, you can make sure your own house is in order by inspecting lighting on your property. Are the tops of bulbs exposed? Are your lights on unnecessarily at an hour when no one is about? Do any lights shine into a neighbor's yard? (She or he may be trying to observe the sky.)

Or try this experiment: Stand under a street lamp. Can you see the bulb from another lamp? If so, the lamps do not have enough shielding. (The light from the second lamp should not be reaching you; it is not required where you stand because you are, after all, standing *beneath* a lamp.)

In this way, you can reduce the light pollution coming from your own home. For sky watching, you may even be able to extinguish all artificial illumination on your property. You can do a great deal to improve your view of the sky by avoiding just a few nearby lights. But ultimately, you have little control over more distant but more numerous lights strung around your city or town.

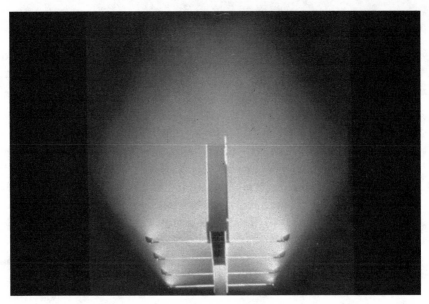

FIGURE 4.4. Light pollution in a modern city. Courtesy of International Dark-Sky Association.

For most people, the best bet for a totally dark sky will be to flee—out of town, or away from any source of nocturnal illumination. Luckily, there still are such places, though citizens of major cities may have a fairly long trip. Regardless, it will be worth the effort. You will be amazed how many stars you can detect at a non-light-polluted site. (An added benefit of this detour is that it may get you away from urban smog, which adversely affects the clarity of the sky.) A convenient test of your site is the Milky Way. If you can spot this band of light, you are seeing the sky as the ancients did—after they had extinguished their camp fires.

If you cannot avoid light completely, find a site that has a relatively dark, unobstructed horizon in the direction of the object you want to look at. Once there, do not introduce your own light pollution. After you are settled, turn your headlights or flashlights off.

If you require a little illumination (for instance, to consult a sky chart), try putting a red filter over your flashlight. Red light does not disturb your night vision as much as other colors do. (This is the same strategy used in lighting airplane cockpits.) As it dark adapts, the human eye becomes less sensitive to red light (in most situations)—a trait called the Purkinje effect. While you can buy special red LCD flashlights, a piece of red cellophane will do. I also have succeeded at making my own red light by painting the bulb of a penlight with red fingernail polish.

Clearly, viewing the sky effectively takes a bit of planning ahead of time. This is time well spent. Why look at a cool constellation from a crummy site? It is like watching the Superbowl from the highest row of bleachers.

Realistically, the final arbiter of your astronomical view will be the weather. There is little that can be done about clouds, besides crossing one's fingers.

Still, assuming the night is clear, when should you look? The simple answer is: Look whenever the celestial object of your choice is above the horizon (the closer to the celestial meridian the better). Remember, though, that twilight must be ended totally (or not yet begun) to have the darkest sky achievable.

If you cannot avoid light pollution entirely, there is a certain practical advantage to observing after midnight. (This assumes, of course, that your target object is above the horizon at this time.) The reason is that more outdoor lighting gets turned off as one heads into the early hours of the morning, and your sky will be a little bit darker.

There are two other factors that will affect your view of faint astronomical objects. First, even if you do the same thing each night, the quality of your sky view can vary. And it has nothing to do with you. The transparency of the air may change from night to night. By "transparency," I do not mean whether it is cloudy or not. Even on a technically clear night, there can be thin material in the atmosphere—material that makes celestial objects appear dimmer or with less contrast. Your best bet, most assuredly, is to observe for as long and on as many nights as possible, to increase your chances of high transparency and a good view.

Second is the moon. The moon is a fascinating astronomical object to study in its own right. However, in some cases (for example, Milky Way watching), moonlight is worse than a bright street light. The moon is a natural source of light pollution, though because it is a "spotlight" pointing down, it is less troublesome than terrestrial lights shining up. The moon goes through its cycle of phases once each month. You cannot do anything about this other than to plan your observing session with the moon in mind. Chapter 8 is about watching the moon. However, it can be read as instructions on how to avoid the moon, too.

EXPECTATIONS

Astronomers toss around words like "spectacular" and "bright." These words are meaningful in the context of astronomical events. But today, movies and television have cranked up the threshold of what people perceive as spectacular. Frankly, we need a lot of sensory stimuli to get excited about an event these days. (You cannot merely have a car crash

Twilight: Last Gleaming?

To better see the stars, it is best to avoid twilight. But when does twilight begin or end? Obviously, dawn ends at sunrise, and dusk begins at sunset. However, when does morning twilight start? When does evening twilight end? These questions are more subjective.

Different definitions of twilight have been used. A Jewish tradition holds that the Sabbath ends once three stars can be made out in the evening sky. Today, three secular definitions of twilight still are used under differing circumstances.

For example, civil twilight is considered to be that time before sunrise or after sunset when there remains enough light to see common objects. In other words, we normally require artificial illumination (i.e., turn the lights on) before morning civil twilight and after evening civil twilight. Formally, civil twilight is defined to exist while the sun's midpoint is between altitude negative six degrees and the horizon. (A negative altitude refers to how far something is below the horizon.) Only the brightest stars are visible during civil twilight.

Sailors use a different standard. After nautical twilight in the evening (or before nautical twilight in the morning), it is no longer possible—absent moonlight—for a mariner to discern the silhouette of the horizon. Sky and sea appear to merge. The formal definition of nautical twilight requires the sun to be at altitude negative twelve degrees or less.

Professional astronomers use an even stricter criterion. They want the sky to be as dark as it is going to get. Astronomical twilight ends earliest (a.m.) and latest (p.m.) out of the three modern kinds of twilight. To avoid astronomical twilight, the sun must be eighteen degrees or more below the horizon. By then, the contribution of the sun to the brightness of the sky is equivalent to or less than that of the stars and the natural faint glow of the air.

Would not the popular vampire-novel series sound cooler if it had been titled *Astronomical Twilight*?

in a 2000s movie; it has to be a whole lot of cars—or at least a bus.) The media bombard us with light and color. This is quite a bit for any natural phenomenon to have to compete with. But it is the very actuality that astronomical phenomena are completely natural events in our universe, putting on their display with no help from—and with indifference to—humankind, that makes them remarkable.

Astronomers inadvertently raise expectations, too. We like to publish our scrapbook pictures of the heavens. Amazing "shots" of distant astronomical objects can be found in books, magazines, and online—in color—all with extraordinary detail. However, such images were built up with sensitive detectors, over long exposures, using huge telescopes. Sometimes they are computer enhanced.

Your view of the same objects almost certainly will be subtle. Remember, though, what you are looking at is real. It is not recorded and edited for your consumption as is so much of what we experience today. You are watching the universe live.

5

The King of Day

Often I have swept backward in imagination six
thousand years, and stood beside our Great Ancestor,
as he gazed for the first time upon the going down of
the Sun. What strange sensations must have swept
through his bewildered mind, as he watched the last
parting ray of the sinking orb, unconscious whether
he should ever see its return.

GENERAL ORMBY MACKNIGHT MITCHEL,
The Planetary and Stellar Worlds, 1848

So much for the stars. Let us "talk" more about the sun. In the daytime
sky, it is the most obvious (perhaps the only) astronomical object visible.
And its light swamps that of everything else combined. The sun's influence on us was obvious from prehistory: Plants and flowers follow it. Animals (and people) change their behavior depending on its presence. All
one has to do is to extend one's palm and feel its life-giving warmth.

The sun appears on the celestial meridian at noon. While the sun
is rising (east of the celestial meridian), we say that it is ante meridiem.
When the sun is setting (west of the celestial meridian), we say that it is
post meridiem. Those of us who do not speak Latin just use "a.m." and
"p.m." The sun defines the morning and afternoon.

Until comparatively recently, astronomers considered the start of
their day to be noon, not midnight, because at noon there is a specific
event at a specific time (the sun on the meridian) that can be noted and
used to calibrate their very exact clocks. There is no such easily observable event twelve hours later, though a midnight transition from one day
to the next is much more convenient for everybody else.

The sun is our biggest timepiece. Recall that the sun appears to slowly change position on the celestial sphere, whereas the stars do not. It makes one circuit in a tropical year. A tropical year (as opposed to a sidereal year) ignores the minor annual effect of precession.

Ironically (because we supposedly base all of our clocks on the sun), sometimes the real sun arrives a bit late to the celestial meridian. Sometimes it is a bit early. This would not happen if the sun moved exactly the same amount and in the same direction on the celestial sphere each day. (Such a tropical day is reckoned with respect to the sun; a sidereal day is reckoned with respect to the stars.) However, the sun moves in different directions on the celestial sphere at different times of year. Moreover, the earth (as well as all the planets) travels in its orbit faster than average when it is closest to the sun, perihelion, and slower than average when it is farthest from the sun, aphelion. (More on the earth's orbit later.) This is a natural outcome of Isaac Newton's theory of gravitation. The result is that the sun appears to speed up and slow down in our sky—at least with respect to our clocks; thus we assume a constant-rate sun. These variations in both direction and speed are cyclical and called the equation of time. The analemma, a strangely shaped figure eight that traditionally appears on a globe, can be read to make the necessary clock correction due to the equation of time. However, days and hours of varying length just will not do. So all earthly timekeeping is based upon a fictitious mean sun that travels at constant angular speed. We will use such a sun as well, in subsequent references to time.

Hours of constant length were not always the case. Before clocks, sun events, like sunrise, noon, and sunset, were used to keep track of time. As an example, the traditional prayer times for Muslims are not governed by the clock. The first prayer is to be made between sunset and the end of dusk. (The Islamic and Jewish days begin at sunset.) The second prayer is to be made after nightfall. The third prayer is to be performed between daybreak and sunrise. The fourth prayer is offered in the early afternoon, when the sun culminates.

ON THE DIAL

Water clocks dry out. Hourglasses empty. Springs and weights wind down. But the sun clock moves perpetually across our sky.

It is not necessary to watch the brilliant sun and the celestial meridian to find the fourth Islamic prayer time (noon). There is a less awkward

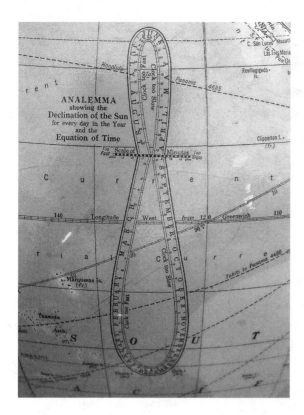

FIGURE 5.1.

The analemma, painted on a globe so as not to obstruct the view of land masses. Photo by Michael Hockey.

method: A stick stuck in the ground so as to cast a shadow is called a gnomon. This most simple instrument is a convenient method of finding real noon. At (local) noon the sun is at its highest place in the sky for that tropical day, so the shadow cast by the gnomon then is at its shortest. (In north temperate latitudes, this shadow will always point north.)

Still, it might be difficult to decide upon the moment of shortest shadow. More conveniently, we can mark the shadow length any time before noon, and when it reaches this same shadow length after noon. As the sun will have been at the same altitude at both times, noon will have occurred at a time halfway between. (Of course, knowing when noon *was* is not the same thing as knowing when noon occurs.)

The fifth Islamic prayer is the most interesting from an astronomical point of view. It is to be made after the gnomon's shadow length has increased by half the length of the gnomon and before it has doubled from its noon length (or the sun has set). Notice that it does not matter what the overall length of this shadow is on any day; that depends on the time of year and latitude—more on this later—and the length of the stick. The relevant quantity is a ratio. The times of day at which these

FIGURE 5.2. The gnomon of a large, modern, public sundial, in Colorado Springs, Colorado. It is over three meters tall. Photo by John Carmichael (designer of the sundial).

relative lengths occur will vary, as a function of season and latitude, but they always will happen.

A gnomon is at the heart of every sundial. Indeed, a sundial is just a gnomon and some sort of plate marked off in hours. The dial may be horizontal or vertical. It is set up so that the shadow of the gnomon falls on the appropriate hour mark at the appropriate time. Of course, sundials keep track of real sun time; they are affected by the equation of time.

Projecting the apparent circular motion of the sun in the sky onto a flat dial, by way of the gnomon tip's shadow, meant that marks on the dial intended to correspond to hours might not be of equal angular separation. The first step toward solving this problem was the tilting of the gnomon to point at an altitude equal to the latitude of the sundial and using the shadow of the gnomon's edge as our indicator. The shadow length was now no longer important; the emphasis fell on the angle at which the shadow is cast upon a radially marked dial. The resulting sundial shadow moves through a constant angle in equal intervals of time during a given day.

The shadow of the old-style gnomon tip fell on a (nearly) unique location for each hour of each day. Using the gnomon edge erases any information about date. Sundials today are foremost clocks, not calendars.

Where Would We Be without Our Stick?

So far our simple stick/gnomon has served us as both a clock and a calendar. It also can act as a daytime compass:

STEP 1: Set up a vertical gnomon.

STEP 2: Mark the tip of the shadow in the morning.

STEP 3: Do so again in the afternoon, when the shadow has reached exactly the same length as when you made your first mark. The exact time of Step 1 is unimportant; it will determine the time of Step 2. (The closer it is to noon when you make your first mark, the shorter the time you will have to wait until making your second mark.)

STEP 4: You now have created two rays, one pointing northwest from the gnomon through your first mark, the other pointing northeast from your gnomon through the second mark. For northern, midlatitude folk, due north is indicated by a ray exactly halfway between the first two. In southern midlatitudes, the third ray points due south.

An alternative to step 4 involves drawing a segment between the two shadow tips of steps 1 and 2. This segment is east-west; north-south is perpendicular to it.

This method assumes that the sun does not move on the celestial sphere during the course of one day. Shortly we will learn that this assumption is not correct. However, the added complication of the sun's apparent motion on the celestial sphere will affect direction determination by gnomon only slightly.

The sundial creates hours of equal duration from sunrise to sunset. Notice that this is not the same thing as hours of equal duration throughout the day. Early in the history of timekeeping, nighttime (between sunset and sunrise) was not necessarily divided into hours at all. Or nighttime hours might be of different lengths than daytime hours. In either of these systems, the length of an "hour" may vary seasonally.

A more elegant version of the sundial, called the heliochronometer, uses a gnomon aligned parallel to the earth's axis, plus a dial that is not flat but, instead, is a segment of a hollow cylinder. The dial is in a plane perpendicular to the gnomon. On a heliochronometer, the angular intervals between hours are of equal separation, and hours are always of equal duration.

The sundial (and heliochronometer) was invented in the Northern Hemisphere. In the Northern Hemisphere, most of the time (exceptions to follow) the sun works its way across our southern sky during the course of the tropical day. Thus, which way will the shadow of the sundial's gnomon appear to turn, from morning to afternoon? Clockwise. The hands of our clocks could just as well turn in the opposite direction. They would function as well. Clockwise (the same direction in which a person screws on a jar lid) was chosen as it is because that was the way a clock was *expected* to move—the same direction a sundial "moves."

Is your sundial off? Even if you have taken into account the equation of time, there is another, systematic difference between the time it depicts and the time you use in your day-to-day life. The sun will reach its noon position in the sky at different absolute times at different longitudes on the earth. The farther west you live, the later noon arrives. That may be of little concern to you: The time intervals between your local sunrise, noon, sunset, and any other such event are unaffected. It gets complicated, though, when people from different longitudes start communicating with each other, or start traveling routinely from place to place. To make things easier, humans have divided the world into twenty-four time zones. The mean sun is on the celestial meridian at noon in the middle of each time zone. It is early in the eastern part of the time zone, and late in the western part of the time zone. However, despite this, everybody living in a given time zone (at least in principle) observes the same time, say eastern standard time.

In theory, the borders of time zones should be arcs of constant longitude. In practice, they have been adjusted here and there, for geopolitical reasons. Some places ignore time zones altogether: The Peoples' Republic of China uses a single, national time across sixty degrees of longitude.

Anyway: When most of the world's people use just twenty-four different times, instead of everybody observing their own local time, confusion is lessened and commerce benefits. The trade-off is that, if you live far from the longitude at which the time in your time zone is set, your sundial may not agree with your watch. The difference will be minor.

What if your sundial is off by an entire hour? The likely blame is daylight saving time. As usual, it is the sundial that is correct. It is our watch that is wrong.

Daylight saving time is an arbitrary advancement of the clock's hands by one hour. This is done in the spring. Clock hands are then returned to their correct position in the fall. The dates and times for making these changes are dictated, not by the sun, but by legislation. These laws vary from place to place and from year to year. Most importantly for us sky watchers, the sun does not "know" we are making the changes.

Daylight saving time is supposed to reduce energy usage. That it does so never has been proven. This notwithstanding, we continue to "spring forward" and "fall back" all over the world. (Before daylight saving time existed, when did people check their smoke-detector batteries?) So if you want to use your sundial for conducting business, do this: When daylight saving time is in use at your location, add one hour to sundial time.

Daylight saving time complicates the simplifying step of introducing time zones. At one time, different parts of Indiana (the state in which I was born) were in different time zones, and the two regions disagreed on whether or not to observe daylight saving time. The result? I was born on a March night near midnight, and there is to this day some confusion as to the date of my birthday. My opinion of daylight saving time is best summed up in this dialogue from *One Day in the Life of Ivan Denisovich* by Alexander Solzhenitsyn (1962):

> A tub was brought in to melt snow for mortar. They heard somebody saying it was twelve o'clock already.
> "It's sure to be twelve," Shukhov announced. "The Sun's over the top already."
> "If it is," the captain retorted, "it's one o'clock, not twelve."
> "How do you make that out?" Shukhov asked in surprise. "The old folk say the Sun is highest at dinnertime."
> "Maybe it was in their day!" the captain snapped back. "Since then it's been decreed that the Sun is highest at one o'clock."
> "Who decreed that?"
> "The Soviet government."
> The captain took off with the handbarrow, but Shukhov wasn't going to argue anyway. As if the Sun would obey their decrees!

FIGURE 5.3. Changing the clocks for spring daylight saving time. Is the hour hand about to be pushed or pulled? Photo by Michael Hockey, and courtesy of Jackson's International Auctioneers & Appraisers.

In summary, to change local sundial time to standard time, you must

1. correct for the time of year (the equation of time);
2. correct for your location in your time zone;
3. (sometimes) correct for daylight saving time.

Ultimately, there is a physical limitation to the precision of the time a sundial provides. Every shadow, produced by an extended object like the sun, has two parts: There is a dark umbra, from within which the light source is completely blocked. Outside the umbra is a softer penumbra, from within which the light source only partially is blocked. The penumbra varies in darkness from that of the umbra (at the penumbra's inside edge), to that of no shadow at all (at its outside edge). (Look at the shadow of your own finger against this page.) Thus, every shadow is characteristically fuzzy. This fuzziness produces uncertainty on the dial. So observing shadows is not a replacement for measuring time by observation of celestial objects themselves.

Eratosthenes Picks Up a Stick

Eratosthenes of Cyrene, who lived around 200 BCE, used a gnomon for another purpose. He used it to measure the size of the earth.

Eratosthenes had learned something interesting about the Egyptian town of Syene (near modern Aswan): On a certain day of the year, a vertical stick, planted in the ground there, cast no shadow at noon. (The sun was at the local zenith.) Yet, at the same time and date in Alexandria (more-or-less due north of Syene), that same stick cast a shadow. (The sun had a lower altitude in Alexandria than at Syene.) Eratosthenes guessed that the sun was far enough away so that its direction is essentially the same at both Syene and Alexandria. So what was the difference between the two sticks?

The difference was in the meaning of the word "vertical." On a spherical planet earth, vertical means "toward the center." So the two sticks were not tipped the same with respect to the sun.

Eratosthenes went one step further than demonstrating that our earth is not flat. On the critical date and time, he measured the Alexandrian shadow length and computed the angle of the sun from the zenith. The difference between this angle and that at Syene (zero) told Eratosthenes what fraction of the earth's circumference lies between Alexandria and Syene (a known distance). There is uncertainty about the Alexandria-Syene distance Eratosthenes used, but his circumference of the earth looks much like the modern value. Not bad for a guy and a stick.

Why are there twenty-four hours in a day? Why not, say, ten? Or one hundred? Twenty-four sounds rather unusual, if you think about it, especially when so many other units are based upon 10^n.

One answer comes from history. We may have the ancient Egyptians to thank for TV's 24 or 48 *Hours* series (as opposed to—say—19 or 42 *Hours*). Moreover, it was because they, like us, were enamored with tens. How does that work?

The Egyptian Sothic[1] year was simplicity itself: twelve thirty-day "months," plus five days at the end of the year to party. (The real moon could do whatever it "wanted"; there was no attempt to match the Sothic year's month with the moon's cycles.) The Egyptian year of 365 days was a reasonable approximation of the true tropical year.

The Sothic year began with the first morning appearance of the brightest star in the sky, Sirius. The Egyptians noticed, though, that after about ten days, Sirius no longer rose to mark the end of night. So they picked another star that would fit this role. And ten days after that, another star. The result was thirty-six stars more or less evenly spaced in a circle around the celestial sphere, each of which rose in the same way, in turn, for about one-thirty-sixth of the year. These special stars were called decans. (Sadly, we no longer know which decans were which stars.)

The utility of decans was that they could be used to tell time at night. For example, as Sirius rose, one would know that the next decan behind it would soon rise, and that the next decan after that would rise after about the same interval of time. Even over the shortest night in Egypt, a minimum of twelve decans would appear sequentially at the eastern horizon. Counting decans permitted the counting of time.

Meanwhile, the Egyptians were marking time between sunrise and sunset in a different manner. You cannot count stars in the daylight. Instead, they used a sundial divided into ten—yes, ten—hours. But what about the time of day during which the sun does not cast the shadow of a gnomon, because it is below the horizon, yet when it is still too light to see decans? The Egyptians solved this problem by adding an hour of twilight to the morning and evening of each day.

So let us tally the Egyptian day up, from dawn to dusk to dawn: one hour of twilight, plus ten hours of daylight, plus twelve hours of night, plus one more hour of twilight, equals twenty-four hours.

Admittedly, the lengths of the twenty-four Egyptian hours varied a little. For instance, each of the daytime hours in the winter was shorter than each of those same hours in the summer. However, in equatorial

Egypt (as we shall see shortly), the difference was not significant. It would likely have affected few people, too: Without artificial lighting, no ancient Egyptian worried about operating his or her business "twenty-four/seven."

THE SUN IN THE LAND OF THE RISING SUN

There are many myths associated with the sky. Most tell of how something came to be. Or why something looks the way it does. What makes the following sky lore stand apart (and a favorite of mine) is that it tries to explain observed motion in the sky.

The *Hagoromo* is an oft-told Japanese tale. It has been translated into children's verse, Noh theater, dance, and epic poetry. Stories similar to *Hagoromo* appear throughout the world.

In one version of *Hagoromo,* a fisherman on the island of Miho comes upon a pool in the forest. It is late afternoon, and seven *tennyo* (celestial maidens) have come to Earth in order to dive and splash in the waters. As they bathe, the *tennyo* have hung their *hagoromo* (robes of swanlike feathers) on trees. The fisherman finds and takes one of the robes. Once the *tennyo* discover the fisherman watching them, they recover their robes and flee to the sky. However, the seventh *tennyo* cannot, as the fisherman has possession of her *hagoromo* and, without it, she may not fly back to the heavens. She pleads with the fisherman for the robe's return. He does not budge. She dances for him in order to entice his compliance. He still refuses. She explains that, as a mortal, the robe is of no use to him. Once again the fisherman refuses and prepares to camp for the night. He doubts the *tennyo*'s supernatural identity. The *tennyo* argues that, if he returns the robe, she will prove who she is by taking flight. As she fails to persuade the fisherman, the *tennyo* grows weak. Eventually, the fisherman notices that the day grows late, but the sun remains in the sky. The shadow the fisherman casts grows no longer. He becomes tired, but evening does not come. Realizing at last what he has done, the fisherman gives the *hagoromo* to the *tennyo*. She ascends to the sky. Quickly, dusk arrives, and the diurnal cycle resumes, propelled by the full complement of *tennyo*.

Hagoromo reinforces the nearly universal cosmology of the separation between the earth and the celestial realm. Bad things happen when beings belonging to one realm find themselves in the other. Moreover, the story acknowledges a common pre-Newtonian physics in which motion must be accounted for by continual impetus—in this case, "angels," who push the celestial bodies along their paths. For these reasons, and because of its popularity, *Hagoromo* is a quintessential sky myth. Yet it remains largely unknown in the West. A 1990 Japanese space probe (to the moon) was named Hagoromo.

THE ECLIPTIC

In time spans greater than a day, it is more obvious that the sun does not stay put like the stars. The annual apparent path of the sun through the sky and stellar background is called the ecliptic. The ecliptic runs through the celestial sphere; the path is centered on us.

In Western tradition, the sun enters, in turn, each of twelve constellations along the ecliptic and then repeats the progression. These are the zodiac constellations famous from the pseudoscience of astrology, and "stars" of that lamest of barroom pickup lines: "Say, what's your sign?"

(Because of the way modern constellation boundaries are drawn by astronomers, the sun spends a brief part of the year passing through Ophiuchus—not a traditional zodiac constellation. The next time somebody asks you what your "sign" is, tell them "Ophiuchus." Try not to get slapped.)

The names for the traditional zodiac constellations, special only because they are the ones the sun happens to travel through as it follows the ecliptic, come from the fact that many of them are supposed to represent animal figures. "Zoo" means animal—as in, well, "zoo"—and "dia" means circle—as in "dial."

The figures of Western zodiac constellations sometimes defy our attempts to match them with existing Greek myths and legends. They are a curious cast of characters. Yet there are some patterns to be seen within them. For instance, those zodiac constellations that once hugged closest to the horizon (in the Northern Hemisphere) can be associated with water: Pisces (the fish), Aquarius (the water bearer), and Capricorn (the sea goat). Were these once the constellations in which the sun hung out during the rainy season?[2]

Here is another familiar word: The intersection of the ecliptic and the eastern horizon is called the horoscopium. Now you know from where the word "horoscope" comes.

What is the significance of horoscopes and the zodiac constellations? If you believe in popular Western astrology—and it is a belief, not a science—your future (indeed, your whole life) is determined by where the sun was on the celestial sphere when you were born. You might think, then, that your "sign," the constellation that the sun is "in" during your birth month, would be displayed prominently at this time each year, your birthday. But, no: Because the sun is "covering up" (in front of) your con-

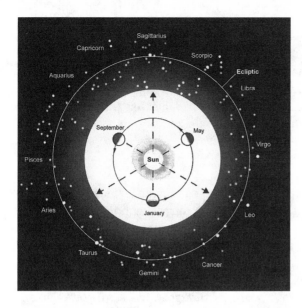

FIGURE 5.4. The ecliptic. The zodiac extends approximately six degrees either side of the ecliptic. Not to scale.

stellation on your birthday, "your" zodiac constellation is the constellation that is hardest to see at this time.

Except that it does not even do this, anymore. The astrologers, when they were making all this stuff up, were unaware of or forgot to do something: They have not kept up. They still are using the locations of the stars on the celestial sphere from thousands of years ago. (Talk about being out-of-date!) Modern tabloid astrologers just keep rereading and rewriting the same old books over and over again, without, apparently, doing any real thinking of their own. The result is that they now are more than one constellation off, due to precession.

My zodiac sign is Aries, coincidentally the constellation the sun used to inhabit at the beginning of spring. Insofar as the year was once considered to start on that date (in Europe and Asia), this point on the ecliptic was long ago significant and given a special name, the First Point of Aries. Yet anyone who takes the trouble to look (perhaps briefly during a total solar eclipse, when the sun's bright disk is temporarily covered) will find that the sun is now in Pisces on my birthday, March 27.

This is not news: While in Aries during the Golden Age of Greece, the First Point of Aries actually has been in Pisces for the last two millennia. (Virgil and others tell us that this transition heralds a change of "Ages"; is the Christian symbol of the fish also meant to reflect a New Age?) These dates eventually will be still one more constellation off when the Age of Aquarius dawns, some centuries hence.

Today's commercial astrologers are not even using the right sun "signs." How strange is that? It is about as strange as the whole idea of trying to tell strangers insights about their own lives, based on lights in the sky. Or insisting that the complexity of all human personalities can be lumped into just twelve kinds. The pop astrologers are going to have to face the unpredictable future, good or bad, just like the rest of us.

OBLIQUITY

Why do we have seasons? What each of us fails to know about the astronomy of the night sky is easily remedied. It is those things we think we know—incorrectly—that cause problems. Misconceptions once lodged in our minds, especially at an early age, are difficult to wedge out.

Why do we have seasons? The answer is not—I repeat, not—because the earth is closer to the sun at one time of the year than another, and thus closer to the source of heat at one time and farther at another. Even though you may have heard (or thought you heard) your elementary school teacher say that this is the case, it just is not true. If that were the correct explanation of the seasons, it would conflict with what we learned from our geography teacher: namely, that while we, in the earth's Northern Hemisphere, are suffering through winter, the folk in the Southern Hemisphere are enjoying summer (and vice versa).

This is a case of a little knowledge being a dangerous thing: Our (imaginary?) misguided teacher recalls that the earth's orbit around the sun is elliptical, with the sun at one of the ellipse's foci. So the earth actually is a bit closer to the sun at one place in its path (say, perihelion) than at another place in its path (say, aphelion). But the earth's orbit is only slightly out-of-round. The difference between closest and farthest from the sun is about 3 percent—not enough to cause the dramatic changing of the seasons we experience.

In an ironic twist for those of us who trudge through northern winters, the earth is slightly closer to the sun now in the month of January. So much for proximity to the sun.

The real reason why we have seasons has to do with something else. Remember the ecliptic? The apparent path of the sun through the celestial sphere? It might have seemed annoying at the time that I was introducing a second imaginary great circle centered upon ourselves, the first being the celestial equator. Will not just one do?

No. The ecliptic is not the same thing as the celestial equator. The former is defined by the earth's revolution about the sun, the latter is defined by the earth's rotation on its axis. The reason for this difference is

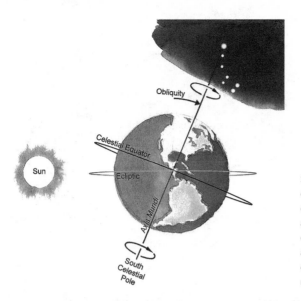

FIGURE 5.5. The obliquity of the ecliptic. Other planets in our solar system rotate on their axes, too, and have their own obliquities. These range from almost zero to more than ninety degrees. The sun and earth are not to scale.

that the axis of the earth is tilted twenty-three and a half degrees from a line perpendicular to the plane of the earth's orbit about the sun. An equivalent way of saying this is that the ecliptic is tilted twenty-three and a half degrees with respect to the celestial equator. This tilt is called the obliquity of the ecliptic. The asymmetry of the analemma, north to south, is due to the earth's obliquity.

The rotation and revolution of the earth are independent of each other. If the celestial equator turned in the plane of the ecliptic, it would be a tremendous coincidence. All the other planets rotate in planes that are tilted with respect to the planes of their orbits about the sun, too.

What does obliquity have to do with our seasons? A lot. Let us think about the earth at different points in its orbit. Regardless of where the earth is, in the contemporary era its axis always points (close to) the direction of the star we call Polaris. Polaris is our North Star year round. The orientation of the earth's axis does not change. Again, revolving and rotating are not the same thing.

You might have thought that, if the earth is moving with respect to the sun, its axis should move with respect to the sun as well. However, this would require some interplanetary giant to hit on the axis with a hammer. During intervals of time with which we are practically concerned, the earth's axis points in the same direction, only changing ever so slightly over the eons, because of precession. This is the same principle as that put to use in a stabilizing gyroscope.

Let us take a close-up look at the earth during our (northern, temper-

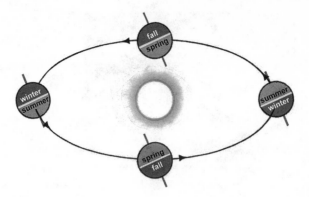

FIGURE 5.6.   North is to the upper left. Imagine the earth orbiting the sun into (at the top of the figure) and out of (at the bottom of the figure) the page. (This perspective might cause the illusion that the earth's orbit is very elliptical; however, if viewed from above the plane of the ecliptic, the earth's orbit would be seen to be nearly circular.) The earth and sun are not to scale.

ate) summer. Our hemisphere is tipped toward the light and heat coming from the sun. Now let us look at the corresponding winter: Our hemisphere is tipped away from the direction of the sun's light and heat.

Returning to the earth and our sky, consider a summer's noon. Remember that noon is the time at which the sun culminates. The sun, as it appears to make its way across the sky that day, is at its highest point at noon. As we glance up at the sun, we must tilt our head far back: The sun is at a significant altitude. It is not at the zenith, but high in the southern sky.

(One could accuse our hypothetical and confused primary school teacher of propagating another myth: For those of us at temperate latitudes or higher, the sun never reaches an altitude of ninety degrees. Not on any noon of the year. So the saying, "The sun is overhead at noon" is, for most people, most of the time, false.)

Now consider the same circumstances in our winter. It is still noon, but this time we need not tilt our head so far back to glimpse the sun. The sun is at a lesser altitude than it was six months ago. In fact, at any hour of a winter's day, the sun is lower than it will be at the corresponding hour of a summer's day.

A key part of this geometric exercise is to demonstrate our own experience when earth hemispheres are interchanged. At the same time and on the same date that we note how high the sun is (compared to a year's average noon), our twin at the same latitude in the Southern Hemisphere comments on how low the sun is (compared to a year's average noon). Except for the fact that she is looking north, "her" sun is in the same place in the sky (the same altitude) that it was for you half

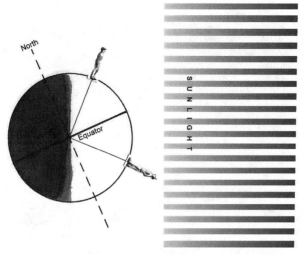

FIGURE 5.7. Two noon sun watchers at similar latitudes north and south of the equator.

a year before (or after). Our summer's day is a winter's day for the twin. The reverse occurs in our winter. The two of us see the same thing (the altitude of the sun), but at opposite times of year—just like in the geography book.

Think of your own experience. Picture that last holiday card you received, the one showing a sunny, outdoors, winter scene. The artist always includes long shadows in the painting—indirect evidence of the sun's low altitude. Now imagine a sunny summer at the seaside: Do long shadows seem appropriate to your recollection? No. Summer is a time when shadows are scarce, even to the degree that we create artificial shade in the form of beach umbrellas.

The obliquity of the ecliptic is not a constant. It varies back and forth about a mean, by a few degrees, over tens of thousands of years. This has implications for the earth's climate. Yet the period and amplitude of the variation is such that it has no practical significance to today's sun watchers.

INTRODUCING A CELESTIAL MAP

All of the above can be illustrated in the form of a map of the celestial sphere. This particular chart will be a cylindrical projection. Mapping the astronomical sky is called uranography.

The proper shape of the earth, a sphere, is best modeled by a globe. But globes can be impractical. Try mounting one on your wall. So for

Dusk & Dawn: How Long?

If the winter sun achieves a lower altitude at culmination than does a summer sun, then it is following a daily path across our sky on average more parallel to the horizon than its counterpart six months hence. So you might think that winter twilights should last longer than those in the summer. However, more important is the fact that the summer sun's diurnal path lies in the vicinity of the horizon, before sunrise and after sunset, longer than does the winter sun's. What is the difference between the sun's behavior at these two different times of year? From the point of view of temperate dwellers the sun must cross from one celestial hemisphere to the other in the summer. Not so in the winter. The result is that, in the summer, the sun spends more time just below the horizon and produces longer dusks and dawns than it does in the winter. In chapter 7, we will see why long twilights are absent altogether near the earth's equator, but standard fare close to the poles.

If your astronomical horizon differs greatly from the real horizon, your dusk and dawn experiences may be out of the ordinary. In a mountain valley, the sun will dip below the surrounding peaks while it is still the brightness and color we usually associate with midday. Sunset (and sunrise) here is a sudden, brief, and even unexpected event.

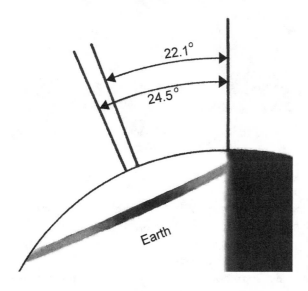

FIGURE 5.8. The obliquity of the earth varies slightly, over a period of time even greater than the Platonic year.

convenience, we "peel" the skin off our globe and flatten it out. When we force the spherical surface of the earth onto a plane, distortions occur: The North and South Poles are no longer points. They take up the entire top and bottom edge of the map, respectively. The earth's equator is no longer a circle. It is a straight line. And, most oddly, instead of being continuous, the earth's surface is interrupted—at the right and left margins of the map. (We have to remember that the left and right sides are in reality the same thing.)

We put up with all of these distortions for convenience, and they soon become second nature. We cease to be offended by the preposterously exaggerated size of Greenland. It is the map we are used to seeing on the wall of nearly ever social studies classroom.

What works for the earth, works for the celestial sphere. Instead of countries, we map stars. We take a globe-like model of the celestial sphere and "peel" its surface. This is flattened out into a map that has all the same distortions as our earth map. The north celestial pole, south celestial pole, and celestial equator are all lines; constellations far north or south are stretched out; and the celestial sphere appears to terminate on either side.

The only difference is that our star map is made to be used looking up, whereas our earth map—think of a road map—is meant to be used looking down. So when north is "up" (its traditional location), east is to the left and west is to the right on the star map. Again, we put up with it all—for convenience.

We are discussing the sun and want to add it to our map. Here we

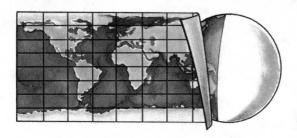

FIGURE 5.9. Making a cylindrical-projection map of the earth.

encounter a problem because the sun is not at one fixed point on the celestial sphere as stars are. The best we can do is to plot its path, the ecliptic, on our new map. The sun appears to circle the celestial sphere once per year, so the ecliptic on our map is a continuous curve all the way from the left-hand side to the right-hand side of the map. The shape of the ecliptic is peculiar: It looks like a wave undulating across the map. But I prepared you for this. You know that the ecliptic is really a great circle. The wavy appearance on our chart is just another map distortion. In order to have a "straight" celestial equator, we must endure a "wiggling" ecliptic, crossing the celestial equator twice.

It is worth it, though. Because the sun is at a given point on our map once each year, the map doubles as a sort of calendar. Three hundred and sixty-five vertical lines across the map would correspond to the days of the year, with the current day corresponding to the line on which the sun sits.

Using our map-as-calendar, certain interesting phenomena pop out at us. For instance, the sun spends approximately half of the year in the north celestial hemisphere and half the year in the south celestial hemisphere. (Equal amounts of the ecliptic are above and below the celestial equator on the map.) It is just those times when the sun is in the south celestial hemisphere that the sun looks particularly low in our northern daytime sky. No wonder: The sun is not in "our" hemisphere. We call these times "wintery." When the sun is in the north celestial hemisphere, and higher in our sky between sunrise and sunset, we call it "summery."

Our Southern Hemisphere friends will disagree. The words "summer" and "winter" mean "warm season" and "cool season," no matter in which hemisphere you live. Our South American, southern African, or Australian neighbors will therefore apply these words six months out of synch with us.

6

Solstices, Equinoxes, and More

At last the sun has reached the cedar spires
That end his pilgrimage upon the hill,
And all the clouds in benediction, now,
Stand robed in flame and for a moment still.

He long has passed the inky-fountained elms,
The black seaweed of oak, the beeches' pale
Fantastic lace . . . And now, the cedars reached
He turns to trace again his ancient trail.

Thus moving like a giant pendulum
Along the tree-fringed hill, to south, to north
He ticks the seasons there eternally
In mighty, golden journeys, back and forth.

FRANCIS MORTON O'NEILL, "Winter Solstice"[1]

THE SOLSTICES

Let us pick up where we left off last chapter. Note that there is a point, on our just-constructed celestial map, where the sun gets as far north as it ever will—this corresponds to the day of the year where the sun is at its maximum noon altitude. That day has a name: It is called the summer solstice in the Northern Hemisphere. This day is considered the first day of summer in the United States—but midsummer in Europe, where the climate is slightly different. It happens on June 20 or 21 each year.

"Solstice" means "sun static." The sun seems to hang at that altitude for a day ("hang time"), then reverse its direction north to south and sink.

Conversely, there is one day of the year when the sun is at its minimum noon altitude. That day is the Northern Hemisphere's winter solstice: December 21 (sometimes December 22). It is the first day of winter or, if you prefer, the day of midwinter.

Do you live in Botswana? Cross out and swap the two dates that appear in the paragraphs above for the Southern Hemisphere's summer solstice and winter solstice.

Many peoples in times of old kept track of the solstices. (The Cahokia sun circle was oriented toward the solstices.) Think how you would react if you saw the sun getting lower and lower each day. The begged question logically would be: Eventually will there be a day when the sun does not make it up at all? Perhaps never to return? This would be calamitous. Therefore, when the sun "turned the corner" and started getting higher in the sky day by day, that event would be a time of great relief and rejoicing. Silly? Not by a long shot. It is not coincidental that Christmas, New Year's, and other holidays are celebrated so near the winter solstice.

How might we determine the solstice date experimentally? Because the sun is not always on the celestial equator, its sunrise azimuth and sunset azimuth change. Only on two days a year is the phrase "the sun rises in the east and sets in the west" strictly true. At all other times, sunrise is a little north or south of east, and sunset is a little north or south of west. (A northeastern sunrise corresponds to a northwestern sunset, at the same angle with respect to an east-west line, each day; the deviation from this rule caused by the apparent motion of the sun during just one day is minuscule.)

The motion of the sunrise or sunset azimuth along the horizon has been likened to that of a pendulum on a grandfather clock. It swings back and forth, centered on due east or due west, from a point in the north (northern-summer solstice) to a point in the south (northern-winter solstice). The farther away from the equator you live, north or south, the wider the pendulum swings. This makes finding the solstice easy to do at high latitude. Just record the date when the sun is at a particular azimuth a few days before the expected solstice. Then record the date on which the sun returns to that azimuth after the solstice. The exact date of the solstice will have been halfway between the two recorded dates.

NEWGRANGE

Where do you suppose the earliest human artifact acknowledging the solstices is? You might be surprised. It is in Ireland, a tomb called Newgrange.

Newgrange today is fifty kilometers north of Dublin. It overlooks the River Boyne as it did when constructed. It has been attributed to the old sky god Dagda. Alternately, it has been attributed to the famous Celts—

but they appeared on history's scene long after the builders of Newgrange were gone. The fact is that Newgrange lay forgotten for most of its history, and was rediscovered only in 1699.

What is Newgrange? It is a mound fourteen meters tall, with a diameter of eighty meters, and is surrounded by a wall of ninety-seven stones. It dwarfs the more-familiar English artifact called Stonehenge; the main monument of Stonehenge could fit easily within Newgrange.

There is an opening in Newgrange marked by an elaborately decorated stone. This door leads into a twenty-meters-long passage that ends in a cruciform chamber six meters high. All the dressed stone work at Newgrange is laid in place (not mortared). There is art on the inside walls, too.

The corbelled ceiling of the end chamber speaks of some complex engineering. There even are drainage channels in the Newgrange stonework to keep water out of the interior. It has been estimated that it would take three hundred men and women twenty years to build a Newgrange.

Newgrange is an example of a passage grave. The bones of some— but not many—bodies still can be found inside. Other, less spectacular, examples of passage graves abound throughout Europe.

A south-facing door automatically suggests some astronomical intent at Newgrange. The sun, moon, and planets all appear in the southern sky for a northern hemispheric observer living this far north. But nothing is meant to shine through the door of Newgrange; stone blocks may have once sealed the entrance shut.

It was archaeologist Michael O'Kelly, hired to restore the monument, who discovered a small "transit" above the door, one meter wide by twenty centimeters tall. It had been plugged by two removable blocks of quartz. Indeed, scratch marks indicate that the block has been removed and reinstalled many times.

Around the time of the winter solstitial sunset at Newgrange, the sun shines through this transit, illuminating the inside of the tomb. It does so for about a quarter hour.

Newgrange was not an observatory. A week either side of the solstice is not good enough for marking the solstice on a calendar. Besides, only the dead would see the phenomenon. Still, Newgrange demonstrates a symbolic connection between human life and the "life" of the sun: old and weak at the winter solstice, vigorous and strong ("at the top of its game") at the summer solstice. The sun will be "reborn" and repeat the cycle. Was it thought that those interred at Newgrange might do so as well?

There are other "satellite" graves near Newgrange that share its orientation. The people who built Newgrange clearly noted and imbued the solstice with significance. And this is the exclamation point: They did so before 3200 BCE. Newgrange is one of the oldest surviving buildings in the world.

Today, Newgrange is a popular tourist attraction. Each year at the solstice, there is a waiting list to get in.

For the rest of us, there is the more recent passage grave in the Orkney Islands, Maes Howe. (It is interesting to call 2500 BCE "more recent.") Maes Howe also is aligned to the winter solstitial sunset and has received some publicity lately because enthusiasts have broadcast, over the Internet, real-time images of the event.

STONEHENGE

There are many alignments around the world to the winter solstice. The same is true for the summer solstice.

When we think of stone monuments, we usually do not think of Newgrange. Stonehenge is the most well-known symbol of archaeoastronomy. There is little doubt that sky watching took place at Stonehenge. There is terrific disagreement over just what sky watching was done.

Stonehenge is located on the Salisbury Plain, in southern England. According to the twelfth-century Welsh monk-turned-historian Geoffrey of Monmouth, Stonehenge was constructed magically by that great sorcerer Merlin. He did so on behalf of King Aurelius Ambrosius, to commemorate the noble knights who fell in a famous battle with the Saxons. Actually, Merlin supposedly stole Stonehenge, or, as it was said to be called originally, the Giant's Dance. We are told that he flew the assembly intact from Ireland. Is this a murky recognition of Newgrange?

Stonehenge previously might have been written about by the Greek historian Diodorus in the first century BCE (though he may have been referring to another stone monument in Britain). Regardless, it was ancient even then. The real Stonehenge builders definitely were not Druids, a common myth. The Druids consorted in England much, much later. Modern folk who call themselves Druids regularly assemble at Stonehenge, but neither they nor the real Druids have or had any real idea of what Stonehenge is all about. The truth is that Stonehenge was built by a culture native to Britain thousands of years before Diodorus.

What is Stonehenge? Stones set upright are sometimes called menhirs. The Stonehenge we see today is the remains of a circle of thirty

Precision

How precise are the astronomical alignments I write about? Would everybody agree on the same direction (even if it is accurate) to within a small angular measurement? Only approximately. After all, the builders of artifacts such as Newgrange were not modern scientists with their modern "obsession" with numbers. Alignments most likely were symbolic; they were not intended as astronomical tools for prediction. Foresights and backsights are not mere points: We do not always know where and how to look; walls and passages are not always perfectly straight; and the objects, or parts of objects, that define the alignment are usually too close to each other to delimit a single, exact direction.

Sometimes claims are made for great archaeoastronomical precision. These usually rely on the use of a distant object as the foresight. Fair enough. However, this distant object probably must then be a natural one, such as a mountain peak on the horizon. If only an artificial foresight is necessary for an orientation, the possibility of accidental alignment increases.

There are other reasons not to seek mathematical precision. We do not always have the artifact as it was when it was constructed. Even massive stones can move over long intervals of time. (One way for this to happen: The ground beneath a Northern Hemisphere rock stays frozen longer on the rock's north side than it does on its south—the rock begins to lean toward the softer south.) People even may cart the stones away.

Newgrange and sites like it were not scientific research institutions. They probably had a religious function. The importance of the alignment is the acknowledgment of the astronomical phenomenon itself.

worked menhirs, five meters high, each weighing at least twenty-five tons. Sitting on top of these were thirty-seven-ton stone lintels. The menhirs and lintels were shaped so that they would fit together like Lego toys.

Within this menhir ring is a horseshoe of five trilithons (like the letter U upside down). Even the heaviest of these Stonehenge building pieces was moved from at least thirty kilometers away.

Going into the past, there is archaeological evidence of at least three separate Stonehenges, one on top of the other. Dates are based on Carbon-14 studies of organic materials associated with the much larger, nonorganic artifacts.

In broad strokes, the historical picture is painted this way: The first Stonehenge was just a 110-meter-diameter ditch. (The word "henge" simply can refer to a circular earthen embankment surrounding a ditch.) It was built around 3000 BCE. There is an opening in the bank to the northeast.

The next Stonehenge included posts of wood and eventually stone. Presently, there is no above-ground evidence for the second Stonehenge.

Finally, by about 2000 BCE, Bronze Age folk added the so-called Sarsen Circle and trilithons. (Sarsen refers to the type of sandstone used.) While there are other stone circles in Britain, only Stonehenge incorporates the technically difficult lintels. Recently, it has been suggested that Stonehenge stones were thought to have healing properties.

The proposal that Stonehenge has something to do with astronomy was first made as early as 1740. All three Stonehenges had gaps in the northeast quadrant. If, in Stonehenge's heyday, you stood in the middle of the stone circle, looking out the opening (and "avenue" leading away from it), you would have seen the sun rise over a more-distant menhir foresight called the Heel Stone. This did not happen on just any day, though. It occurred on one special day: the summer solstice. (Admittedly, you could have looked 180 degrees around and watched the winter solstitial sunset, too, half a year later.) It is noteworthy that both the lone Heel Stone external to the circle and the opening that begins the Stonehenge "avenue" date from the first Stonehenge.

What do I mean exactly when I say that Stonehenge marks the summer solstitial sunrise? Unfortunately, the sun does not rise on solstice day in quite the same direction now, as it did in 3000 BCE. This is because of the exceedingly slow but measureable change in the earth's obliquity. Now the sun appears a couple of degrees left of the Heel Stone. By the time the entire disk of the sun appears, it is still a degree away from the Heel Stone. Yet when the sun is high enough to clear the Heel Stone, the sun

FIGURE 6.1. The Heel Stone at Stonehenge. There are varying accounts as to how it got this odd name. Photo by Chris Collyer (www.stone-circles.org.uk).

is right over it. Evidence from excavation indicates that something once stood near the Heel Stone, but to the left of it. Maybe the now-missing stone and the Heel Stone formed a pair, one in which the rising sun was neatly framed?

Many other astronomical alignments have been claimed for Stonehenge. The solstice is by far the most robust.

Author Bernard Cornwell, better known for Napoleonic War novels, has written *Stonehenge: 2000 B.C.* (2000). In it, he spins a tale in which Stonehenge, as we see it today, was assembled within the lifetime of a single person (the protagonist). This scenario almost certainly is wrong. However, it is fun to watch how Cornwell works around the archeological evidence. The novel is not about astronomy, but it is fun.

ANTICIPATION

If you want to celebrate an event such as a solstice, you need to be able to anticipate that event. You need time to prepare. Planning and scheduling remain important in regard to the ceremonies practiced by Native Americans. Today, wall calendars are free at pharmacies, and our watches and cell phones provide the current date at a glance. But what about before those tools existed?

One of many cities built by the Ancestral Pueblo people of the Four Corners region of the United States (the intersection of Utah, Colorado, New Mexico, and Arizona) is located at a place called Hovenweep. Now a National Monument headquartered in southeastern Utah, it was constructed around 1200. Inside one D-shaped building, called The Castle, holes only ten centimeters in width pierce thick walls. Such holes are not good windows, at least as far as ventilation is concerned. Moreover, some of the holes are angled with respect to the walls that hold them. Why was this done unless the holes were meant to point toward something?

Archaoastronomer Ray Williamson thinks that the room in which we find these holes was an observing room. In it, a person could sit each late afternoon, near the time of solstice, watching the light from the setting sun move across an interior wall—in anticipation of the solstice. Did the light's position on the wall provide a "count down" to the date of a ceremony? The time to put the crops in? There is a different hole for each solstice. Were there once special markers or statues on these walls? In another Hovenweep building, called Unit Type House, the same sort of architecture can be seen.

Modern Puebloans, the descendants of the *Ahi-sat-si-nam* (as they call their ancestors), practice certain agriculturally related rituals at solstice time. They use the azimuth of the sun to remind themselves when it is time to prepare. So anticipating the sun at Hovenweep is plausible: American astronomy practiced since just short of a thousand years ago.

Ancestral Pueblo rock art can be found all about Hovenweep. One example has been chipped into a rock face such that a spear of light, caused by landscape shadows, pierces a spiral sun symbol shortly after dawn on and near the summer solstice. I have watched it—it was worth the effort of getting up early! The interplay of natural light and shadow is just one more way people have discovered in which to participate in celestial events.

INSOLATION

Admittedly, it is not the altitude of the sun that we commonly think of as marking the difference between the seasons. For most of us, when it is winter, it gets cold. When it is summer, it gets hot. *That* is what we notice. Why the change in temperature?

This turns out to have everything to do with the altitude of the sun.

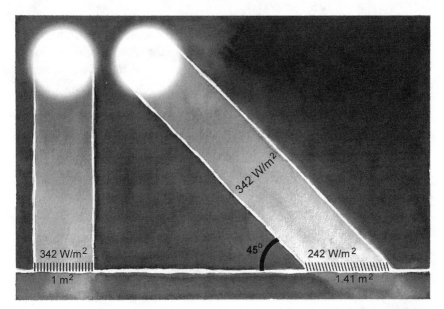

342 W/m²

342 W/m²

45°

242 W/m²

1 m²

1.41 m²

FIGURE 6.2. Seasonal insolation at two different times of year. The units used in this diagram stand for "watts per square meter."

For it is the sun that provides nearly all the heat the surface of the earth receives. Imagine a bundle of the sun's heating rays. We can do this because there is only a finite amount of solar heat intercepted by the earth at any time. ("Solar" is a prefix meaning "of the sun"; it comes from a Latin proper noun for the sun: Sol.) For simplicity, let us have it be midday. When the sun is low in altitude, the bundle of energy intercepts the earth obliquely. That amount of energy is spread over a large area. Any one square foot, yard, or meter does not get much of it, and there is not much warming. After all, when we warm our hands at a campfire, we hold them out flat, not obliquely.

Think of solar heat as paint from a bucket. If we splash it out onto the floor at a narrow angle, a lot of floor gets painted, but at no place is the paint very thick. If we dump the bucket nearly straight down, all the paint is deposited thickly in one (albeit small) place. You will not have to worry about giving this spot a second coat.

Back to the earth and sun. Meanwhile, in the other hemisphere, the sun is higher in the sky, and an equal bundle of heat hitting the earth here is not spread over such a great area. There is little dilution. The heating is more concentrated, and more effective. The heating efficiency of the sun is called insolation.

Our illustration took place at noon, but all through the day the insolation will be more efficient where the sun's energy strikes the earth closer to the vertical. The briefer time the sun appears above the horizon in winter further reduces insolation during that time of year (while the longer daytime hours of summer add to solar heating then). Whereas these effects are most dramatic on a solstice, they apply, to a lesser and lesser extent, for days a quarter of a year before or after the solstice, too.

While the particular weather we experience, at a particular location on the earth, varies from date to date, and year to year, most of us associate one particular time of year as that during which it is hottest and another time of year during which it is coldest—on average. These dates normally are not the solstices, and we would not expect them to be.

How do you make tea? Do you place a kettle of tap water on the stove, immediately remove it from the burner, and then soak a tea bag? I do not think so. The result would be lukewarm tea at best. It takes time to heat up a whole kettle of water.

It takes even longer to heat up (or cool down) an entire hemisphere of the earth. There is a time lag between maximum insolation and the date on which our hemisphere is warmest. This lag may be six weeks long. Similarly, it takes time to cool an entire hemisphere of the earth. Again there is a time lag (of similar length) between minimum insolation and the date on which our hemisphere is coolest.

Where I live, we wish that the coldest day of winter was in late December. In fact, these days usually are mild compared to what we experience near the beginning of February. Likewise, our worst summer heat is in July or August, not June.

In the United States today, we hear a lot about designing energy-efficient buildings. The concept is old news, though: Such architecture was practiced in the American Southwest a thousand years ago by the Ancestral Puebloan people.

The biggest settlement built by the Ancestral Pueblo was in Chaco Canyon, New Mexico. We know the Chacoans paid attention to the sky. For instance, one of their great *kivas* (sunken ceremonial structures) may mark significant astronomical events. Casa Rinconada (circa 1100), as it is now called, is twenty meters in diameter. Inside, there are twenty-eight niches placed around the wall. A twenty-ninth may have been lost in reconstruction. It looks as if the sun was meant to shine in upon something (statuary?) in the niches at special times of year, for example, around the summer solstitial sunrise. (Reconstruction makes it unclear whether now unseen walls, or roof supports, might have interfered with the line of sight; they probably did.)

The Chacoans appear to have put their sky knowledge to practical purpose: The building now known as Pueblo Bonito (also circa 1100) could house perhaps 1,200 people at one time. It is four stories high and divided into about six hundred "apartments." The overall plan takes good advantage of solar heating in the winter by symmetrical arrangement about south. Terracing allowed heat absorbed by the front rooms during winter days to be reradiated to the back rooms at night. Moreover, the design and location provide plenty of shade to common spaces in the summer. Add to this the thick insulating walls, the mass of which helped to equilibrate extreme temperatures, and you have all the hallmarks of a green building.

THE EQUINOXES

So much for the solstices. I now point out two other special locations/ dates on our celestial map. Notice that there are two places—two times of year—when the sun is not north or south of the celestial equator, it is *on* the celestial equator. These dates are about halfway between the solstices. When the sun appears to be moving southward and crosses into the south celestial hemisphere, we mark the occasion, calling the day the autumnal equinox, the first day of autumn to some. This happens on September 22 or 23. The day when the sun moves back from the south celestial hemisphere to the north celestial hemisphere is, of course, the spring (or vernal) equinox—northerners' first day of spring. It is March 20 or 21. (The dates of solstices and equinoxes may vary by a day on our modern calendar, depending upon whether it is a leap year or not, for instance.)

How do we remember which equinox is which? "Spring up," and "Fall down." It is puerile, but it works.

The sundial (or any) gnomon does something interesting on the equinoxes: Its shadow at sunrise and shadow at sunset lie in the same line. Only on the equinoxes is the sun on the horizon at opposite azimuths. (This is true in either hemisphere.)

The world "equinox" comes from Latin words similar to "equal nights." And it is true that, just like a star on the celestial equator, the sun at an equinox spends twelve hours above the horizon and twelve below— almost.

On weather broadcasts, the meteorologist often includes sunrise and sunset times for the day. You may have noticed that, even on the equinox dates, the sunrise time is not just the sunset time with the "a.m." and "p.m." switched. Who is right? The weatherperson or I?

Both of us. The sun is not a point, but a disk. When we talk about sunrise and sunset, are we talking about the geometrical center of the sun? No, because when the middle of the sun is at the horizon, there still is plenty of light from the half of the disk above the horizon.

The edge of a sphere appearing as a disk is called its limb. We are more likely to think of sunrise as the moment the sun's top limb, the vertex, first emerges from below the horizon. We are likely to count as daytime any time when any part of the sun is above the horizon. So, we will not call it sunset until the sun's vertex sinks below the horizon. Our definitions of sunrise and sunset artificially lengthen the daytime. Yet that is not the main factor in the overly long equinox daytime.

The earth's atmosphere acts as a natural prism, bending the light entering it downward from space. This atmospheric refraction is zero at the zenith, and at its maximum at the horizon. At low altitudes, the positions of astronomical objects are apparently shifted by refraction because the geometric direction to the object and the direction from which the light comes are different. Uranographers have to take atmospheric refraction into account, but most of the time its effect is too small to be noticeable to the rest of us.

The case of the sun is an exception. Because we can see the sun all the way down to the horizon, where there is appreciable refraction, the last rays of the sun shine on us *after the real sun already has set*. Its light is bent up over the horizon. Sunset is delayed, and sunrise comes early. The result is that the length of daytime is extended by several minutes.

One last word on the four special dates I have discussed in this chapter: Though their significance is unknown to me, there are automobile models on the road right now named the Solstice and the Equinox.

THE BIG HOUSE

The city of Phoenix, Arizona, is so named because it is built over the remains of an older, indigenous civilization. They were the Hohokam, "Those Who Have Gone," in the tongue of present-day Tohono O'odham. The Hohokam are most remembered for their hundreds of kilometers of irrigation canals built in the Gila River Valley.

In the center of one particular Hohokam village was a "great house," named (appropriately enough by later folk) Casa Grande. The structure, built in the fourteenth century, still stands. It is made out of layers of mud nearly one meter thick. Wood beams help support its three and a half stories. Casa Grande is an eighteen-by-twelve-meter rectangle, 10.5 meters tall. The structure survives the culture that built it, which died

FIGURE 6.3. This photograph was *not* taken on the equinox. Photo by Phil Plait.

Egg on Their Faces

Can you balance an egg on the equinox? I hardly know where to start with this one. Sometimes I can come up with a piece of history or culture that explains the origin of such strange bromides. Here, though, I am at a loss. It is usually a broadcast weather reporter, with air time to kill, who propagates the myth of balancing eggs upright on the equinox. Nonsense. Of course, this is true: With a steady hand, you can balance an egg on its end—on this date *or on any other date*. (Eggs are not perfectly smooth; I find that secretly crushing the shell a bit at the base "helps.") There is no physics—no force—that applies at, or differently on, equinox dates that does not also apply on every other day of the year.

out sometime after 1400. Catastrophic floods, drought, and soil salinization have been blamed for the collapse of the Hohokam's overextended agricultural system.

Casa Grande was one of the first archaeological sites preserved by the U.S. government. It has since been stabilized (different from reconstruction); for instance, a seemingly out-of-place roof of metal now shades it from rain and sun.

The perpendicular walls of Casa Grande were aligned cardinally. An opening (which looks to us like a porthole) is angled toward the summer solstitial sunset. Moreover, similar east and west portals point toward the equinoctial sunrise and sunset, respectively. (Remember that the sun rises/sets in the same direction on both the vernal and autumnal equinox—due east and west.) The equinoctial orientation is crude, but if intended, it means that the Hohokam were concerned with at least three of the sun's inflection dates.

An equinoctial alignment is more difficult to make than is a solstitial alignment. Why? While near a solstice, when the sun's apparent velocity along the ecliptic is changing from northward to southward (or southward to northward), its average speed is zero. The Hohokam (or anybody else) would have several days to record the direction of the solstitial azimuth, on each of which the sun is at approximately its solstitial angle from the celestial equator. This is handy, especially because cloudy weather could easily wipe out several days of sunrise or sunset sightings.

However, the opposite is true at equinoxes. The sun's azimuthal speed is greatest here. (On our star map, the slope of the ecliptic is steepest near an equinox; it is flat on a solstice.) Each day before and after the equinox, the sun is significantly away from the equinoctial azimuth, north or south of the celestial equator. Effectively, you have one day to get the equinox right. To see the same equinox again, "wait 'til next year."

The approximate equinoctial orientation at Casa Grande simply may fall out of aligning an orthogonal building to the cardinal directions. It may have nothing to do explicitly with the equinoxes. (This is always the dilemma with purported equinoctial orientations, though it must be said that it is not easy to establish east and west exactly—even if you know north and south—without *something* to mark these directions.) If they are intended, they reveal a certain sophistication in astronomical technique among the Hohokam. And we have reason to believe that these people may have possessed this sophistication: There is plenty of evidence that the Hohokam came from or traded with older Mesoamerican civilizations to the south. There is a cultural bridge all the way from Teotihuacán to Casa Grande.

CROSS-QUARTER DAYS

Solstices and equinoxes—are there any other special days during the year of the sun?

To reiterate: In the United States, summer is said to begin on June 21; in the United Kingdom, however (with its somewhat different climate), that date is considered to be midsummer. Remember Shakespeare?

So if the solstices mark midsummer and midwinter, when do summer and winter begin? It is possible to further divide the year by introducing those days halfway between the equinoxes and solstices. These are called cross-quarter days. There is a cross-quarter day approximately six weeks after the vernal equinox (six weeks before the summer solstice), six weeks after the summer solstice (six weeks before the autumnal equinox), six weeks after the autumnal equinox (six weeks before the winter solstice), and six weeks after the winter solstice (six weeks before the vernal equinox). The British summer, for instance, would then begin at the cross-quarter day before the solstice and end on the cross-quarter day after it.

(The word "approximately" occurred in the last paragraph because the sun does not reach each of the solstices, equinoxes, and cross-quarter days in equal intervals of time. This is because the earth does not travel at the same speed in its orbit about the sun, year around. Presently, the asymmetry is increasing. Centuries ago, the season lengths nearly were the same. The variation is cyclical, and—in the distant future—the seasons' lengths once again will be equal.)

But has anybody ever paid attention to cross-quarter days?

A segue. Thousands of years ago, the inhabitants of the British Isles erected stone rings across the countryside—Stonehenge is the most famous example. The rings may be only approximately circular; sometimes the menhirs do not form a ring at all. Still, these artifacts clearly are projects requiring skill and cooperative strength to complete. (The stones may not be indigenous to the location of the ring—these may have been imported from afar.) As stone monuments represent most of what has been left to us by this intriguing people, theirs is called popularly a "megalithic culture."

Asymmetric stone rings often demonstrate a solstitial orientation. However, in looking for alignments, it is not always clear which menhirs are to be used as foresights, which menhirs are to be used as backsights, and which backsights refer to distant foresights on the horizon. There are so many potential stone pairs that orientations may easily pop up due to chance alone.

It is difficult to assign with high confidence specific alignments to individual rings. However, there are so many megalithic rings across Britain and Ireland, as well as mainland Europe, that it becomes possible to study them statistically.

Summer after summer, British engineer Alexander Thom (1894–1985) surveyed multiple orientations found in ring after ring. His work was taken up, and made more statistically rigorous, by Clive Ruggles. Eventually they had an impressive number of orientations, which they plotted on a graph of frequency versus azimuth. If all the orientations were random, one would expect the plot to be more or less flat. It was not. There were peaks at azimuths that corresponded to special days (with troughs in between). These special days were the solstices and equinoxes and also, with lower-level statistical confidence, the cross-quarter days.

Unlike on the equinoxes and solstices, the sun undergoes no obvious change in its motion on cross-quarter days. So are cross-quarter-day alignments to be believed? We would be more confident if we had more evidence. Such evidence exists.

The Celtic peoples of northwestern Europe came after those of the ring-building culture. The Celts were practically modern compared to the megalithics. However, we have knowledge of the "Celtic calendar" that has been passed down to today. (I put "Celtic calendar" in quotes because there is debate as to whether there ever was one single calendar or, for that matter, one unified group called the Celtic people; "Celtic calendar" is a simplification.) If we imagine the Celts as our link between prehistoric people and modern times, Celtic traditions suggest practices that might have taken place in their own prehistory.

The Celts indeed did mark the cross-quarter days—as holidays. And by cultural analogy, so may have the megalithics. Below is a table of Celtic holidays and their meanings, listed next to solstices, equinoxes, and cross-quarter days. Also included are Latin and contemporary holidays near which these Celtic days fell. (The Roman Church often incorporated indigenous holidays into its calendar, but gave them different meanings.)

1/8 YEAR	ASTRONOMICAL	"CELTIC"	LATIN
3/21	Vernal equinox		Easter/Passover
5/6		Beltane [move cattle]	May Day
6/21	Summer solstice		"Midsummer"/ Feast of St. John
8/6		Lughnasadh [1st fruits]	Lammas/ Loaf Mass[2]
9/22	Autumnal equinox		Yom Kippur/ Rosh Hashanah/ Arbor Day?
11/5		Samhain [New Year]	Halloween/ All Saints' Day
12/21	Winter solstice		Martinmas/ Christmas/New Year's
2/4		Imbolc [lamb season]	Candlemas/ Groundhog Day

You may be surprised by some of my inclusions in the last column. Nonetheless, Easter is officially tied to the vernal equinox, and Christmas does not occur on a likely date for the birth of Jesus as described in the Bible—but rather, near the winter solstice, as does New Year's Day.

More colloquial holidays fall on cross-quarter days. Of course, Halloween, May Day, and the day that has metamorphosed (in the United States) into Groundhog Day are all very old. (Arbor Day is not an ancient holiday; I throw it in for fun.)

Coincidence? Probably not. Do our holiday chains of electric lights have provenance with pagan, solstitial bonfires, burning to chase away the long darkness?

FIGURE 6.4. The Greek god Atlas holds the celestial sphere in this Roman statue. Shown is the intersection of the ecliptic and equinoctial colure, in the constellation of Aries (the ram). Photo by E. C. Krupp (Griffith Observatory).

CONCLUSION: MAP AS CALENDAR

Returning one last time to our celestial map/calendar: The seasons easily are marked on it. Just as the celestial sphere can be divided in two, north/south, by the celestial equator, an imaginary circle through the celestial poles divides the celestial sphere east/west. There are an infinite number of such circles; but by convention, a uranographer marks the one that also goes through the equinoctial points and the one that also goes through the solstitial points. These are named the equinoctial colure and the solstitial colure, respectively. On our flat map/calendar, the colures are vertical lines dividing the map into four sections. (Each colure appears twice.) One section is where the sun resides in spring; the next, in summer; the next, in autumn; and the next, in winter.

If we morph our calendar/map back into a celestial sphere, the celestial sphere may be represented by just the great circles of note upon it. Such a model in physical form is called an armillary sphere. An armillary is a set of intersecting circles that includes the celestial equator and ecliptic, and (often) the equinoctial colure and solstitial colure. Measurements in the sky could be made using an armillary with scales on its rings. However, the armillary was more likely to be used as an astronomical teaching tool, though it has since gone out of fashion.

Yet, in Portugal, the link between the sky, navigation, and trade is not forgotten. Portugal became a modern nation on the decks of its sailing ships. An armillary is part of the state seal even today. Portugal is the only country I know that uses an astronomical instrument for its symbol.

7

Around the World with the Sun

Late lies the wintry sun a-bed,
A frosty, fiery sleepy-head;
Blinks but an hour or two; and then,
A blood-red orange, sets again.

ROBERT LOUIS STEVENSON,
"Winter-Time," in A Child's Garden of Verses, 1885

So far I have mentioned two things that distinguish summer from winter: the observed maximum altitude of the sun, and the degree of insolation. Can you think of a third? The length of daytime (period of daylight) is shorter in the winter than in the summer. It is the altitude of the sun and length of daylight that cause deciduous trees to change leaf color, not temperature.

DAYTIME AND NIGHTTIME

Just as a star in the south celestial hemisphere spends little time above our horizon here in the north, so does the sun when it is in the south celestial hemisphere. It rises late and sets early. The result is the short daylight hours of winter (and long nighttime hours). In our northern summer, when the sun is in the north celestial hemisphere, it spends more than half a day above the horizon. We enjoy the long daytime for summer recreation.

Any sphere placed in a beam of light is 50 percent illuminated. The rest of the sphere is in its own shadow. The earth in sunlight is no exception: Half the world is always in daylight, half is not. (I am ignoring atmospheric refraction for the moment.) As the earth rotates, a given location on the earth may move from daytime to nighttime. The circle separating the two is called the terminator (a word that has taken on additional meaning since Arnold Schwarzenegger got hold of it). It is the sunrise/sunset demarcation, all around the earth.

Any terrestrial location travels in a circle of constant latitude, due to diurnal rotation. Because the earth turns at a constant speed, once around per twenty-four hours, the length of the arc in daylight determines the length of time that any place on that circle spends in daylight. The more of the arc that is lit, the longer daytime is experienced at the location.

Think of the earth with its North Pole tilted toward the sun. It is the northern-summer solstice. For any location we choose in the midlatitudes of the Northern Hemisphere, most of our latitude circle is illuminated on this solstitial day; we experience the longer "days" of summer. Yet, at that same latitude in the Southern Hemisphere, the opposite happens. Most of the latitude circle is not illuminated; denizens of those locations experience the short "days" (and long nights) of winter.

As the year goes by, the geometry of the earth and sun changes. Eventually the earth's North Pole no longer points toward the sun. It points at a ninety-degree angle to that direction. (The earth has traveled one-quarter of its annual journey about the sun since the solstice.) It is the northern-autumnal equinox. The earth's axis is not pointing toward or away from the sun. Half our midlatitude circle is illuminated, half is not. Equivalently, the length of the sunlit arc through which we travel on that day is equal in length to that of the shadowed arc. On this day, half our time (day) is daytime, half is nighttime.

In fact, it does not matter where we are on the earth at the equinox. The terminator passes over both poles. No latitude is affected differently from any other latitude. Wherever we are, we spend half of our day in daylight, and half in the dark.

The earth has now made its way halfway around the sun since we began watching it at the northern-summer solstice. Now, at the northern-winter solstice, the illuminated arc lengths in the two hemispheres are reversed. Ours in the Northern Hemisphere is shorter than that in the Southern Hemisphere. Thus, ignoring equation-of-time effects, the length of daylight and nighttime in either hemisphere is the same as it was in its counterpart half a year before, complete with the symmetric reversal we associate with a mirror image.

When the earth reaches the northern-vernal equinox (three-quarters of a year from when we started), the day/night experience the world over is the same as it was at the autumnal equinox. And after that it is once again the northern-summer solstice one year after we began.

I am addressing the duration of daytime and nighttime. The actual hour, minute, and second of sunrise and of sunset is governed by the equation of time. The equation of time is a reflection of the earth's variable orbital velocity at different times of year, but it also is affected by the earth's obliquity. The asymmetrical equation of time assures that the dates of earliest sunset and latest sunrise will occur near, but not necessarily on, the winter solstice. Nor will the dates of latest sunset or earliest sunrise necessarily occur on the summer solstice.

Incidentally: If you live at the equator, the preceding discussion might be met with a yawn. There, it does not matter what time of year it is. Because of the symmetry of the globe, the sunrise and sunset hours do not vary at the equator—it is as if it were an equinox all the time. Indeed, seasonal effects are all muted at the equator.

THE ARCTIC[1] AND ANTARCTIC CIRCLES

On the other hand, seasonal effects become more extreme as we travel closer and closer to either of the earth's poles. For instance, we end up with some long winter nights in the far north. Eventually, we come to a latitude at which the length of northern-winter solstitial nighttime equals twenty-four hours. Have you ever wondered what the significance of the Arctic Circle is? It is the latitude for which on at least one day, the sun does not rise at all: 66 1/2° N (again, ignoring atmospheric refraction). Notice that this latitude is just ninety degrees minus the obliquity of the earth. Farther north, there are more of these sunless days on either side of the solstice.

High-latitude residents get these missing daylight hours back: At the Arctic Circle, on the northern-summer solstice, the sun does not set at all. Twenty-four hours of sunlight is often an excuse for a big party in these lands. The tourist brochures call it the Land of the Midnight Sun, ignoring the fact that, six months before and after, it is also the Land of the Midday Gloom. Again, if you go farther north, there are more days of continuous sunlight.

Saint Petersburg, Russia, while not as far north as the Arctic Circle, experiences a summer string of days on which the sun never sinks so far below the horizon that twilight ceases. These are the so-called White Nights.

The effect on people of a seemingly ever-present sun is a theme in the suspense movie *Insomnia* (2002), starring Al Pacino, Robin Williams, and Hilary Swank. Continuous night, on the other hand, is overstated. While the sun may not rise above the horizon, it may still cause twilight. There is no inhabited place on the earth that does not experience at least a bit of dusk/dawn on every day of the year. (At high latitudes, the sun is so close to the horizon throughout the day that, when there is twilight, it is always long, year round.)

Of course, if you are an international jet-setter, once the sun sets at your far-northern latitude, you could fly off to the same latitude in the Southern Hemisphere and never have to watch the sun go down. There is an Antarctic Circle, too (66 1/2° S).

Things get really weird near the North and South Poles. Of course, because the altitude of the sun never gets very high up there, it is always cold. Moreover, as the number of sunset-less or sunrise-less days increases (while we increase in latitude), we eventually reach 90° S or N. The earth is still spinning, but as we stand on the axis about which it turns, *we* are not going anywhere. The earth's rotation no longer affects the altitude of the sun. The sun just circles the horizon over the course of the day, maintaining nearly constant altitude. The only change in altitude that takes place does so over the longer period of the earth's revolution. That revolution carries us pole sitters into and out of the earth's shadow once per year. The day and year have become one and the same: six months of darkness and six months of daylight. Beware of the job advertisement for a South Pole station that reads "only need to stay over one night."

THE TROPICS

There are two other circles drawn on the globe that catch our attention. These are labeled the Tropic of Cancer (Northern Hemisphere) and Tropic of Capricorn (Southern Hemisphere). The fact that each is twenty-three and a half degrees from the equator hints at something to do with the sun. These are the latitudes where the sun is at the zenith on one solstice. As the sun is always within twenty-three and a half degrees of the celestial equator, it is only within the Tropics that the sun can be directly overhead. With the sun always within twenty-three and a half degrees of the zenith at noon, it is always high in the tropical sky, thus explaining tropical warmth.

(As an astronomer, I pride myself on being able to tell which direction is which, at least when the sky is clear. However, once upon a time I found myself in an unfamiliar Mexican city at noon in June. It so happens that this city is near the Tropic of Cancer. The sun was at the one place in the sky where it was of no use to me in telling north from south from east from west: the zenith. I had to resort to my least favorite navigational technique—asking passersby for directions.[2])

North of the Tropic of Cancer, the sun always appeared in our

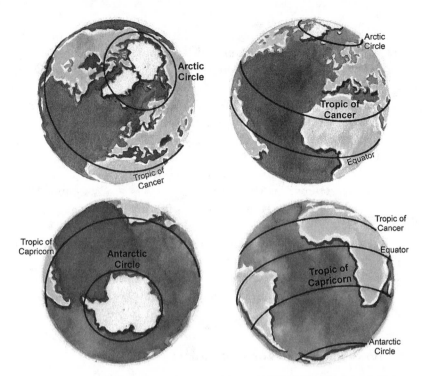

FIGURE 7.1. *A,* The Arctic Circle (*left*); The Tropic of Cancer, 23.5° N (*right*). *B,* The Antarctic Circle (*left*); The Tropic of Capricorn, 23.5° S (*right*).

southern sky, no matter what time of year. Sometimes it was higher than average at noon, sometimes it was lower. But it was always south. South of the Tropic of Capricorn, the sun is always in our northern sky. South and north of the Tropics, respectively, it can cross the equinoctial that divides our sky into northern and southern halves.

The simplest example is on the equator itself. With the celestial equator passing through the zenith and dividing the sky north-south, the sun, which spends nearly equal amounts of time in each celestial hemisphere, will spend nearly equal intervals in the northern and southern skies. The dates upon which the sun passes from the south to the north (and north to south) are the same as the days it passes from one celestial hemisphere to the other: the equinoxes. On these two days, the sun appears (to those of us squatting on the equator) at the zenith.

(In the Tropics, celestial action is often high in the sky. This is good, because the jungle frequently found in tropical lands obscures the view of events near the horizon.)

The sundial plays interesting tricks at the equator. Imagine that it is the equinox. The gnomon's shadow appears at sunrise, aimed due west.

It points that way all morning, continuously shortening until it disappears at noon. It then grows again, this time pointing east until sunset. The shadow tip traces a straight line throughout the day, never "dialing" at all.

Assuming that we are in the Tropics, though not on the equator, there still will be two days on which the sun passes from the northern sky to the southern sky and vice versa, and two days upon which it will reach the zenith. However, these dates are neither solstices nor equinoxes.

Dusk and dawn are brief affairs in the Tropics. The tropical sun escapes the horizon quickly. The diurnal path of the sun is so tipped with respect to the horizon that there is only a short interval of twilight at any time of the year.

Did you notice that the Tropic of Cancer and Tropic of Capricorn are misnamed? Once upon a time, the sun did reach its solstitial destination on the celestial sphere when the sun was in the constellation of Cancer or Capricornus. Precession since has changed that. Still, once something gets painted on the globe, it is hard to change its name.

LIVING NEAR THE TROPIC OF CANCER

In the Oaxaca Valley of Mexico, a couple of hundred kilometers from Teotihuacán, is the ancient abandoned city of Monte Albán. It was built by the Zapotec people, who were contemporaries of the Teotihuacános.

In planned cities like this one, buildings out of alignment with the plan are always something for which to look. In Monte Albán, there is such a place called (fairly unimaginatively) Building J. Building J catches one's eye right away: It is oddly shaped and seemingly askew to the arrangement of other Monte Albán buildings. The shape resembles an arrow. In fact, one side of the building nearly forms a right triangle, but no two sides of Building J are of equal length. Odd geometry is another thing to look out for when searching for astronomical significance.

Is the arrow pointing at something? Yes, but not in the way the word "arrow" feeds our preconceptions: When Building J was built, in about 500 BCE, the back of the building—exactly opposite the arrow—pointed to the rising point on the horizon of the yellowish star Capella (alpha in the constellation of Auriga). This means that if you stood on the steps of Building J, you would look at that point on the horizon where Capella appeared (depending on whether or not you could see it close to the horizon).

Capella is the sixth brightest star in the sky. Sixth place? We need more than that to establish that Capella was special at Monte Albán. By

Finding Your
Place in the Sun

If you know the day of the year, you can use the sun's location to determine your latitude. The date will establish how many degrees north or south of the celestial equator the sun is. The altitude of the sun can then be converted to altitude of the celestial pole (equal to latitude) with simple arithmetic.

This technique is handy at sea. In the daytime, you may not want to await dark in order to determine your ship's position. You may have run aground by then.

A mariner's astrolabe works well for navigating by the sun. This device consists of a circular plate, marked off in degrees. Attached to the middle of the plate is the midpoint of an arm that pivots. At one end of the arm, and perpendicular to it, is a small occulting disk. A tiny hole pierces the disk. There is no hole in the similar disk at the other end of the arm. The navigator holds the mariner's astrolabe by a cord or chain attached to the rim of the plate. This allows the astrolabe to hang normally, under its own weight, even on a pitching deck. Top is always top, and bottom is always bottom. The navigator then points the astrolabe's arm toward the sun, rotating it so that sunlight passes through the hole in the (upper) occulting disk. The result is that a tiny image of the sun is projected onto the other (lower) occulting disk. The direction in which the arm now aims, the altitude of the sun, is read off the plate. With no optical parts made of glass, the mariner's astrolabe was popular until the invention of the more-well-known sextant.

coincidence, Capella performs a special show on or near the day when the summer sun passes through the zenith. (This is a "heliacal rising," to be discussed later.)

Remember that the zenith is probably going to be more important than the solstices to people in the Tropics.

The Cappella event could have been used as a signal that one of the zenith-transit days was at hand.

But what then? Suppose you made the requisite observation of Capella one night. What did you do the next day? You (or someone designated or privileged to do so) may have gone to another building designed, not for observing objects on the horizon, but rather objects at the zenith.

Just next door to Building J is Building P. As you stand on the "porch" of Building J, you are looking in the direction of Capella rising—and also in the direction of Building P. Here, archaeologist Horst Hartung found that the sun shines down a tunnel in the ceiling of Building P at noon on the two zenith-passage days. Such a vertical tunnel is called a zenith tube. The Building P tube is so narrow that sunshine reaches the floor only on these two days (plus maybe a day before or after). Further excavation at Building P will show whether the opening in question is unique, thereby strengthening the argument that it was intended as a zenith tube.

In ancient Greece, they might have used a shadow-casting gnomon to note this kind of zenith phenomenon, not a well. However, the Mesoamerican tradition is much more like that of the Egyptian pyramid contractors.

Incidentally, fifty kilometers from Monte Albán is another contemporary city, built by the same people. It is home to the only other building found to date to be shaped like Building J. It is oriented similarly with respect to another building in the way that Building J is to Building P. Is it a copy—one in which the significance of the original may have been lost?

Long, long ago, did anybody actually know there was such a place as the Tropic of Cancer (or Tropic of Capricorn)? That is, did they realize that there was a place where the sun appeared at the zenith just once per year, and on a solstitial date?

There is evidence that our friends, the Teotihuacános, searched for such a place and found it. Five hundred kilometers north of Teotihuacán, archaeologist J. Charles Kelly excavated a ruined city today named Alta Vista. He found an inhabitation modeled after Teotihuacán, complete with a Temple of the Sun and pecked crosses. There is enough astronomical symbolism and alignments to suggest that Alta Vista was a city interested in the sky.

Alta Vista sits within two kilometers of the Tropic of Cancer—pretty close. At the solstice, the sun's behavior may have established the location of this city.

Founding a city based on its artificial proximity to an imaginary circle, and not with regard to natural geography (nearby water, easy defense, etc.)—that is crazy, is it not?

Maybe, maybe not. The modern city of Laughlin, Nevada, was established so as to be near the imaginary lines that form the "bottom corner" of Nevada. While gambling is mostly legal in Nevada, it is more restricted in neighboring states. Laughlin's location puts its casinos nearer to the well-populated cities of Arizona and Southern California than does its famous competitor, Las Vegas. So which "artificial" location is crazier?

Tiny Necker Island is the last islet in the Hawaiian chain worthy of the name. It is too small for human habitation, yet it is the site of thirty-four temple platforms.[3] What is special about Necker? Could it be that it sits smack on the Tropic of Cancer? Or was it of existential interest merely for being the end of the line?

LIVING NEAR THE TROPIC OF CAPRICORN

High in the Andes Mountains of what is now Ecuador, Peru, and Chile, the people we call Incas united under strong leaders and carved out an empire in less than a century. The capital city of the Inca Empire was Cuzco, once populated by one hundred thousand people (out of the millions of citizens within the empire).

Little remains of ancient Cuzco since the Spanish conquest. Fortunately, we at least have extensive descriptions of it from Spanish chroniclers. The plan of the city fascinated the Spanish because it was unlike any other city they knew—Cuzco was not laid out on a grid, and the plan of Cuzco included a radial component. At the city's traditional center was the Coricancha. Today, the Coricancha is also called the Temple of the Ancestors or Temple of the Sun. The Coricancha originally was covered in gold—before the Spanish pried it all off.

From the Coricancha radiated out conceptual lines (not always straight) that divided the city and surrounding lands into forty-one pie-shaped sections, each one administered by a different family or clan. These lines are called *ceques*. Some *ceques* defined larger areas than others, presumably allotted based upon political clout. This sort of division made geographical sense because Cuzco lies in a mountain valley. Within each "piece of the pie" there was a great variety of elevations, representing all the ecozones the Incas needed to farm and graze—often all within a short (horizontal) distance of each other.

Which *ceque* line went where? We do not know how these decisions were made, but astronomy played a part. Apparently the Coricancha was (among other things) one big backsight. Natural horizon features, or constructed monuments, far away were foresights. Inca archaeoastronomers Anthony Aveni and R. Thomas Zuidema have found *ceque* alignments to the sun and stars at special times—such as the solstices. Indeed, the circular pattern of Cuzco's *ceques* may be used as a calendar. Incan agriculture had many important days: There were different planting and harvesting dates appropriate for different crops suitable to different elevations.

One high-value orientation is so because a tower on a horizon hill marked the foresight. Cuzco is north of the Tropic of Capricorn, so the sun passes through the zenith there. However, this particular alignment points to the sunset on the day the sun passes, not through the zenith, but through the nadir (August 18).

What a strange alignment! Nevertheless, on the Incan calendar, this August day was the traditional start of the planting season. (Remember, we are in the Southern Hemisphere, but not terribly far from the equator.) Nothing of the various horizon towers remains today. The Spanish likely pulled them apart, stone by stone, to build their new cities.

Does a nadir alignment seem too difficult? The Incans were serious about their architecture and their astronomy: The most beautiful of Incan cities was not Cuzco, but Machu Picchu (sort of a king's summer retreat town). Machu Picchu just barely fits into a nook in the mountains. There is no room for horizon-to-horizon *ceque* lines here. Instead, there is a building that has windows through which one can observe important celestial events. It is similar in construction to the Coricancha, and it is aligned to the June solstitial sunrise. The angular precision of this alignment is the best the naked eye can do and is possible because of the use of narrow sighting-tube-like windows. Such precision came at the cost of cutting through more than two meters of wall. The Incas were willing to sweat for their astronomy.

AND FARTHER SOUTH

In the language of the Mocoví, native peoples of Argentina, the words for "north" and for "high" are the same.[4] This sounds odd in the Southern Hemisphere, where the south celestial pole is "high"—in the south. However, this is just one example of where we northerners need to leave our preconceptions at home in our own hemisphere. For the Mocoví, the sun and moon are at their daily greatest altitude in the north. The linguistic association makes perfect sense as soon as we start thinking "Southern Hemisphere."

LIVING IN-BETWEEN

In today's mathematical culture, it seems reasonable to divide the horizon into four sectors, of equal number of azimuths, that we might call the north, the south, the east, and the west. However, if you live between the Tropic of Cancer and the Arctic Circle (or Tropic of Capricorn and the Antarctic Circle), there is another way to divide the horizon that made just as much sense to indigenous peoples living there.

There is a sector—let us call it "the east"—in which the sun rises. Its limiting azimuths are the most northerly and southerly points on the horizon at which the sun emerges, sometime during the year. (These azimuths occur on the solstices.) Likewise, there is a sector—let us call it "the west"—in which the sun sets. Its limiting azimuths are the most northerly and southerly points on the horizon at which the sun disappears, sometime during the year. (These azimuths occur on the solstices, too.) Then there is a sector—let us call it "the south" ("the north" in the Southern Hemisphere)—in which the sun never rises nor sets, but instead, traverses above the horizon. Finally, there is a sector—let us call it "the north" ("the south" in the Southern Hemisphere)—in which the sun never appears. Period.

The angular sizes of these sectors vary with latitude: "The east" and "the west" take up more and more of the horizon as one gets closer to the Arctic (or Antarctic) Circle. "The north" and "the south" take up more and more of the horizon as one gets closer to the Tropic. Yet, for a given latitude, these practical definitions of "the north," "the south," "the east," and "the west" make as much or more sense as arbitrary sectors of ninety degrees each.

HELIACAL RISING

Let us now consider the stars and sun together. So far, the words I have used to describe astronomical phenomena likely sound familiar: rising, setting, and even culmination. The term "heliacal," which has come up a couple of times, probably does not. This is ironic, because in ancient, but still historical, times, the meaning of "heliacal rising" would have been commonplace knowledge. This is a concept that has dropped out of our popular, cultural usage.

The word "heliacal" obviously has something to do with the sun.

FIGURE 7.2. This diagram recreates the heliacal rising of a star. Each picture represents the arrangement of the sun and star, weeks apart, near daybreak. They differ due to the apparent annual motion of the sun. In the first, earliest picture (A), the star does not rise heliacally. Nor does it in the second picture (B). However, by the third picture (C), the star rises heliacally.

The prefix "helio" is a Greek term referring to the sun. The heliacal rising of a star occurs in the early morning of the day on which a star can first be seen in the remaining hours of twilight. Previous to that date, the star was lost in the glare of the sun. Before that, it was an evening star that rose after the sun and was invisible the morning.

Imagine watching morning after morning for a favorite bright star. Then, one special morning, you can see it—but only for an instant because soon thereafter, as the sun gets higher in the sky, twilight gives way to daylight and the star (like all stars in the daytime) disappears from view. You have witnessed the heliacal rising of a star. A star that rises heliacally appears to pop out of nowhere, and then to vanish just as quickly.

This is a noteworthy event. While we may never have heard of a heliacal rising before, there are plenty of written records that document ancient peoples the world over paying attention to the heliacal rising of stars and planets. The significance of the Pleiades at Teotihuacán was that they rose heliacally on the local date of solar zenith transit. At Monte Albán, Capella rose heliacally, also on the local date of solar zenith transit. The Egyptian year started with the heliacal rise of Sirius; for millennia it happened to mark the time of the annual Nile River flood.

A star that heliacally sets does so just after sunset.

Note that it is difficult to predict the exact date of heliacal rise or heliacal set. Depending upon the brightness of the star, these events are affected by atmospheric conditions and the presence of moonlight.

The Dog Days of Summer. Everybody knows that this expression refers to the hottest, muggiest time of year. But what do dogs have to do with the seasons? In the days of Ancient Rome, Sirius, nicknamed the Dog Star, rose heliacally around the beginning of August, the *dies caniculares* (days of the dog). Precession has since put an end to the coincidence, but we still use the Greek phrase. The bank robbery that inspired the Al Pacino movie, *Dog Day Afternoon* (1975), took place on August 22, 1972.

The opposite of a heliacal rise is an acronical one, that is, the star rises—not just before sunrise but also just after sunset (on the opposite side of the sky). It can set just before sunrise, too, for that matter; that is "acronical setting." Notice, though, that acronical events are not as easily defined as are heliacal ones: Exactly when is "just before" sunrise and "just after" sunset? Collectively, heliacal and acronical risings and settings are called stellar phases.

AN EXAMPLE FROM NORTH AMERICA

Just as the Ancestral Puebloans were likely influenced by the earlier cultures south of them, those we call the Native Americans of the plains could have been influenced by the pueblo dwellers to their south. (By "plains," I mean that area between the Mississippi River and the Rocky Mountains.) This cultural influence can be surmised to include astronomical practices. The people of the plains were at most one hundred thousand, spread over millions of square kilometers. Sadly, these people left no writing and no indestructible architecture.

Stones are one of the few construction materials available on the dry steppe that would survive time. When nonindigenous settlers arrived on the plains, they discovered dozens of large stone rings. These rings are not to be confused with the much more common circles of stones that once held down teepees: The bigger rings to which I refer also had "spokes" of stones radiating from the center. Some of the spokes, as well as the centers, were marked by cairns containing, in some cases, tons of rock. They were named "medicine wheels," for no other reason than that they suggested something mysterious; nobody knew why they were built the way they were.

The medicine wheels appear in unlikely places: One example is the twenty-eight-spoke Bighorn Medicine Wheel, located on a flat area high in the mountains (nearly three thousand meters elevation), but having a clear view of the horizon. (It is near modern Sheridan, Wyoming.) From it you easily can see for a hundred kilometers.

There can be snow all year round at Bighorn. My students and I trudged through drifts up to my waist in order to reach the site in May, and we were the first to make it that year. It is not a very hospitable place to reach. (Interestingly, though, mountain winds tend to keep the Bighorn Medicine Wheel itself free from accumulated snow.) Are Bighorn and its counterparts just abstract art? If so, it is an odd gallery. Artists tend to "hang" their work where it can be readily seen, do they not?

Solar astronomer Jack Eddy heard of the Bighorn Medicine Wheel in the 1970s. He visited it and found it to be a thirty-meter-diameter "circle" with a central cairn four meters in diameter and one meter high. Eddy looked at recently available areal photographs—such photographs give you the best perspective on the place. He determined that five of the six spokes having significant cairns at their ends were astronomically oriented.

Two seemed to be aligned to the summer solstice. (One of these is a particularly long spoke.) One alignment is for sunrise, the other for sunset. If you missed solstice sunrise due to clouds, perhaps you could observe solstice sunset?

Other major radial spokes at Bighorn are oriented to stars: the heliacal rising of named stars Aldebaran, Rigel, and Sirius. They are among the brightest in the sky, and roughly evenly spaced around the celestial sphere.

About three hundred years ago, when the Bighorn Medicine Wheel is thought to have been built, Aldebaran happened to rise heliacally at the time of the summer solstice; about a month after Aldebaran, Rigel rose heliacally; and another month later, Sirius rose heliacally. None of these alignments are particularly accurate, especially when atmospheric refraction is taken into account.

Remember how many spokes there are in the Bighorn Medicine Wheel? Twenty-eight. Twenty-eight is the number of days on which the moon is visible during the month. Were the extra spokes on the Bighorn Medicine Wheel—the ones without terminal cairns or alignments—added later, merely to get the total number up to twenty-eight?

While Bighorn is accessible only just before the summer solstice, two months after the solstice marks the end of the time when this site can be inhabited safely. The Bighorn Medicine Wheel, and its location, suggest some sort of shamanic site. (It is not big enough for, nor can it support, a great many people.) Once you arrive for the solstice, the star alignments tell you when to leave.

Eddy went on to investigate other medicine wheels.

The wheel that Bighorn most resembles can be found in Saskatchewan, Canada, near Regina. The central cairn of the Moose Mountain Medicine Wheel alone holds eighty tons of rock. The summer solstice and star alignments are the same as at Bighorn. However, Moose Mountain is dated two thousand years earlier. Was Bighorn a copy of Moose Mountain?

There are enough medicine wheels, and so many potential alignments, that we cannot be sure that they are not the product of chance. Certainly the fact that many medicine wheels exhibit no alignments weakens the astronomical argument for them. Only Bighorn and Moose Mountain appear to be robust examples.

For what purpose were the other medicine wheels constructed? Maybe they were simply navigational markers on the otherwise featureless plain. Maybe they just pointed toward other medicine wheels.

We regret that we cannot simply ask somebody about the history and function of the medicine wheels. Were they the sites of the famous vision quests we hear about in Native American culture? Was the Plains Indian meeting lodge modeled after the medicine wheel? (There is a superficial resemblance.)

We would like to ask all these questions. After all, there still are Native Americans living in the American and Canadian West. Yet, sadly, inquiries about the medicine wheels have run into a dead end. Though the sites may be revered, nobody seems to truly recall anything about the wheels' original purpose.

If we have the dates of the sites right, we could imagine people passing down this information to succeeding generations during the almost two millennia between Bighorn and Moose Mountain. Now they have forgotten—how fragile is human knowledge without writing.

※ (

As we take leave of the seasons, a final reflection: Today's urbanites are increasingly removed from a direct relationship with the seasons. Our definitions of the seasons may have devolved to "sets of three months, beginning with December, March, June, and September, respectively." However, read this Latin translation provided by classicist Daryn Lehoux. It is from Marcus Terentius Varro's (116–27 BCE) *De Re Rustica*.

> [The year is] divided in eight parts: the first from the (beginning of the) west wind to the vernal equinox, forty-five days; from there to the rising of the Pleiades, forty-four days; from this to the solstice, forty-eight days; thence to the sign of the Dog Star, twenty-seven days; from there to the autumn equinox, sixty-seven days; then to the setting of the Pleiades, thirty-two days; from which to the winter solstice fifty-seven days; then to the (beginning of the) west wind, forty-five days.[5]

Eight seasons. Not only that, but interpreting their beginnings and ends requires knowledge of solstices and equinoxes, stellar phases, and weather. How much more sophisticated than four, three-month intervals starting on December 1, on March 1, on June 1, and on September 1!

8

Many Moons

"Never mind the blue poodles," said the King.
"What I want now is the moon."

JAMES THURBER, *Many Moons*, 1943

While the sun was in paramount regard to most societies—old and new, the moon often was elevated to the status of a deity second only to the sun. It is easy to see why: The moon is the second-brightest object in the sky by far ("number two" only to the sun). Indeed, it is often easily seen in the daytime. (Those who say you must wait for nightfall to see the moon likely have not looked for the moon during the day.) Besides the sun, the moon is the only other regularly appearing object in the sky displaying a disk shape.

Still, sky watchers usually assigned a character to the moon different from the personification of the sun. The moon is inconstant (i.e., it has phases; it does not always look the same—as the sun pretty much does). Moreover, it is not uniform in appearance: Even a casual observer may notice dark smudges on its disk. And you cannot hold your hand up to the moon to feel its warmth, as you can toward the sun. It is a "cold hearted orb that rules the night," sang the Moody Blues, back in the day.[1]

These properties gave the moon a somewhat mysterious quality. The moon was sometimes considered a female god, likely because the period of its phases is close to that of the human menstruation cycle.[2] Elsewhere, it was associated with moisture as its connection with earth tides was recognized. Sometimes it was both.

I previously mentioned the misconceptions from the past that we harbor in our minds. Another common one involves the mechanism for the moon's phases. They do not require that shadows be cast on the moon by other objects. (That is something else—an eclipse.) In fact, phases are simpler than this.

THE MONTH

Like the sun, the moon is a wanderer. The moon is a satellite of the earth. It can be seen to move in its orbit, against the background of the celestial sphere. (For those who feel that the moon ought to accompany every night sky, it may come as a surprise that the moon is ordinarily only seen above the horizon for roughly half the twenty-four-hour day.) Furthermore, this motion easily is noticed: The moon moves by an amount equal to its own diameter, with respect to the stars, in only an hour—more quickly than any other periodic celestial object. While the times from star-rise to the next star-rise and sunrise to the next sunrise differ only by minutes, we might see the moonrise lag an hour or more on successive nights, as the moon moves eastward in its orbit. This nightly delay in the moon's appearance is called retardation.

The month is the orbital period of the moon. The word "month" comes by way of "moon." Without the moon, our twelve-month calendar (twelve signs of the zodiac, etc.) probably would not exist. The month is a handy unit of time, much longer than that other celestially defined unit—the day, but shorter than the similarly defined unit of the year.

Yet there are different ways to technically define the month. Perhaps the most obvious is one we have used to mark a time interval before: with respect to the reference point of a star on the celestial sphere. Imagine that we start our stopwatch as the moon passes over a certain star. (Or, more generally, is at a certain angular position on the celestial sphere.) We wait until the moon once again occupies that celestial spot, and then we stop our watch. The length of time we have just measured is called the sidereal month. It is equal to 27.3 days.

The sidereal month is a straightforward way to measure the revolution of our satellite. What could be easier? Nonetheless, 27.3 days does not sound like the months we are used to. We expect thirty-one or thirty days (or at least twenty-eight or twenty-nine in February) per month. This is because our Western calendar is not based on the sidereal month.

The Chinese (for longer than anyone knows) have divided the celestial sphere into twenty-eight *hsiu* (lunar lodges, or sometimes, mansions); the Chinese moon occupies, on average, one mansion per day. However, this is not quite the same thing as a sidereal-month calendar. The lunar mansions were arranged around the celestial equator, and not the ecliptic, which the path of the moon better approximates. Most cultures do not keep track of the sidereal month.

But never say "never": The Borana, nomads who inhabit Ethiopia and Kenya, have a twenty-seven-day calendar in which each day corre-

Parallax

So far, we have acted as if the earth is a mathematical point and that the objects in the sky are points as well. In other words, we treated bodies, including the earth, as if they had no physical extent (no diameter). This simplified geometry works most of the time, when the distances between bodies are far greater than the physical sizes of the objects. It does not work so well in the case of the moon. The moon is close enough to the earth that parallax can be noticeable.

Parallax is the apparent shift in location of an object, with respect to a distant background, when viewed from two different places. Parallax provides us with our depth perception. (We view the world through two eyes separated by several centimeters.) Until recently, it was the basis for all modern surveying.

The moon may appear in a slightly different position with respect to the distant stars, depending on where you are located as you observe it. At moonrise and moonset you may be as much as six thousand kilometers away from, and perpendicular to, the line between the centers of the earth and moon. The moon may appear to shift its location with respect to the stars when viewed sequentially from these two locations in space. Similarly, an observer at the North Pole sees the moon in a slightly different direction (with respect to the stars) than an observer at the South Pole.

FIGURE 8.1. Normally it is difficult to make the bright moon and the stellar background appear on the same photograph. Often it is hard to even see stars near the moon. The reason is that the moon is so much brighter than the stars. They are lost in the moon's glare. Here, a photographic composite shows the moon and stars together on the celestial sphere. Photo by T. A. Rector and I. P. Dell'Antonio/NOAO/AURA/NSF.

sponds with the moon's relationship to a star. Such a calendar is particularly well suited to those who live near the equator insofar as celestial objects rise and set nearly perpendicular to the horizon at low latitudes. When the moon rises or sets alongside one of the Borana calendar stars, the satellite and star are north and south of each other, and form a line parallel to the horizon.[3]

In an even more complex nod to the sidereal month, the New World Maya may have kept track of the lunar version of heliacal star-rise. In Mayan texts, we read that Moon Woman sometimes is preceded by a herald. Such a leading character has been equated to, not an individual star, but a constellations of stars. Such a constellation would rise just before the moon. At other times, Moon Woman is accompanied by a spouse; they walk side by side. If the "spouse" is a constellation, the moon and it rise together. Finally, Moon Woman may carry a "burden," beings who ride on her back. If we again interpret such a character or characters as a constellation, it rises just after the moon.[4] The concept of a "spouse" or "burden" would be meaningless if we were discussing the sun, because the constellation would be invisible as, and especially after, the sun rises. I know of no equivalent to moon-linked constellations like these elsewhere. If we are reading them right, the Maya certainly imagined "outside of the box" that is Old World astronomical thought. Do not worry: We will spend more time with the fascinating Maya later in the book.

Beside the sidereal month, another way to measure the moon's period is to use as a reference point that other hard-to-ignore source of light in the sky—the sun. Suppose that the moon is a certain angular distance away from the sun. The period between this occasion and the next time the moon has the same relationship with the sun also is called a month.

But the sun appears to move, too. Measuring the month is like timing a race run on a circular track. The starting line is also the finish line. This heat, though, somebody picks up the start/finish line while the race is being run, and moves it farther around the track. (The sun and moon appear to move in the same direction on the celestial sphere.) When they reach the place from which they started, the runners must continue onward to break the tape. The duration of the race is necessarily longer.

Because the year is roughly twelve months long, the sun will advance about one-twelfth of the way around the ecliptic in the time it takes the moon to circumnavigate the celestial sphere once. Our new month is one twelfth of a sidereal month longer—about two days. We have measured the synodic month.

The synodic month is 29.5 days long. Half days are inconvenient, so we round this number up to thirty or thirty-one days (with one exceptional month).

MOON PHASES

Why do we commonly use the synodic month rather than the sidereal month? It is because the most unusual "behavior" of the moon repeats over the interval of one synodic month. I refer to the phases of the moon. It is the lunar phases, after all, that we pay most attention to in timekeeping. ("Lunar" equals "of the moon.") Not only is the period from one phase to its next appearance a readily recognizable one, the continuously changing face of the moon with time makes it easier to subdivide the

What's Wrong With February?

EXCEPT FOR February, the calendar month is always longer than the synodic month. This difference, called the monthly epact, is half a day in thirty-day calendar months and one and a half days in thirty-one-day calendar months. But February is shorter than a synodic month.

You might wonder what calendar makers have against February. That is the month that "gets the short end of the stick" so as to make the number of months fit (to the nearest day, at any rate) into a 365-day year. It need not be February. However, even though the earth's trip along its orbit is continuous, we have to start our calendar year somewhere. In the Western calendar many of us use, the year originally began with March. Re-member March is the month during which the sun travels from south to north across the celestial equator, across the First Point of Aries, so this is a sensible time at which to begin the year (at least in the Northern Hemisphere). February was "short changed" without prejudice—it was just the month that happened to occur when the number of days in the year ran out. No offense, February. Personally, I think a truncated February makes us northerners *believe* that the winter is somehow shorter than it is—an added psychological benefit.

Here is French astronomy-popularizer Camille Flammarion's (1842–1925) list of the months in their original Latin order:

MARTIUS (after the god Mars)
APRILIS (after the word for "to open")
MAIUS (after the goddess Maia)
JUNIUS (after the goddess Juno)
QUINTILIS (meaning "fifth"; *Julius* Caesar took this month over for himself)
SEXTILLIS (meaning "sixth"; not to be outdone, *Augustus* Caesar claimed this month)
SEPTEMBER ("seventh"; subsequent Caesars were less successful at co-opting months)
OCTOBER ("eighth")*
NOVEMBER ("ninth")
DECEMBER ("tenth," as in the *decimal* system)
JANUARIUS (after the god Janus)
FEBRUARIUS (after the god Februus)

Thus we start with the god of war (always important to the Romans) and end with the god associated with honoring the dead.

*We sometimes still use the Latin words for numbers. When an American woman gave birth to eight babies, the tabloids nicknamed her the "octo-mom."

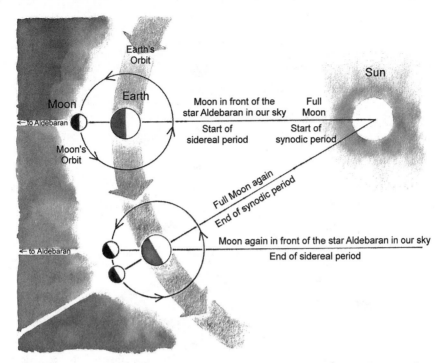

FIGURE 8.2. The difference between the sidereal month and the synodic month. The earth, moon, and sun are not to scale.

month into other units, such as weeks. If the position of the sun in the sky represents the hand of a twenty-four-hour clock, the different phases of the moon represent the hand on our synodic-month clock.

Unlike the sun, the moon produces no light of its own. The moon shines by reflecting light from the sun. (This is the same reason why any of us shine.) As is the case for all spheres placed reasonably far away from a light source, half of the moon's surface is illuminated at all times. The other half is self-shadowed. So, just like the earth, one hemisphere of the moon is in daylight while in the other it is nighttime—always.

Here is the important part, the part that addresses the potential misconception stated earlier: How much of the day-lit moon we can see from here on the earth determines the phase of the moon. No shadows.

Take any sphere, say an apple. Hold it in the light of a lamp or other light source. In this model, you play the role of the earth. Move the apple/moon around your head; just do not let it pass into your head's shadow. You can recreate every phase of the moon that you have ever seen in this simple manner. (It is OK to turn your head in order to see the phases.) Moreover, they will occur in exactly the same order as they do in the real sky.

Long Days

The analogy between the earth day and sunlight falling on the moon is a good one because, in fact, the moon rotates on its axis in one month, too. (More on this in chapter 10.) So the "day" *on* the moon is one month long: two weeks of daylight and two weeks of darkness, on average. In the lunar sky, the *earth* rises and sets, exhibiting phases over the course of the month.

In the iconic motion picture *2001: A Space Odyssey* (1968), a lunar excavation exposes an artificial-looking monolith. This dig takes place just after lunar nightfall, thereby allowing a full two weeks for an Earth scientist to be selected to investigate, his voyage to the moon, and his excursion to the remote excavation site, all before lunar night is over. At daybreak, its first in ages, the now uncovered monolith emits a piercing radio signal in the direction of the planet Jupiter—the beginning of a strange adventure there. However, I think much of all of this was lost on the moviegoing audience. I read the book.

FIGURE 8.3. The phases of the moon. The inner circle of small moons represents what an astronaut high above the earth-moon plane would see at different times of the month. Notice that half of the moon is in sunlight and half not, at all times. The outer circle of moons represents what an observer on the earth would see at those same times. The earth and moon are not to scale.

Traditionally, the synodic month begins with new moon. The sun, moon, and earth are lined up, in that order. An alignment of the sun, earth, and moon is called a syzygy, to the delight of Scrabble game fans.

Remember, the number of days in each calendar month is wrong; the 30-31-30-31 ... pattern is out of synch with the real synodic month. So new moon will not occur on the first of every calendar month. Any lunar phase can (and will, eventually) fall on any day of the calendar month, given that enough years have passed by. Another way of saying this is that there need not be a full moon at Halloween—contrary to the opinion of most TV show writers.

In truth, there are really 365.24 days in a year. Because the synodic month is commensurate (divides evenly into) nineteen years of 365.24 days, a given phase of the moon will reoccur on the same calendar date at most only once every nineteen years (235 synodic months). This interval is called the Metonic cycle.

When, exactly, does new moon occur? When the moon is near the sun, it is lost in the glare of the much brighter sun. Many people think that new moon occurs the evening on which the moon becomes visible again—instead it is the instance when the moon is closest to the sun in our sky and, thus, is invisible to us.

The moon may appear absent very briefly from our sky. The age of the moon is the elapsed time since the moment of new moon. Observations by practiced observers have reported ages of less than twenty-four hours—the record is fifteen—at the first naked-eye sighting of the moon. First sighting is important because it is still used to establish the dates of certain religious holidays.

The time during which the moon is visible to us can be called the lunation. (More standard definitions of "lunation" describe it as lasting from new moon to new moon, but we already have a term for that: synodic month.) A couple of days after new moon, the moon is far enough from the sun in our sky that most people will be able to distinguish it and agree that the start of a new lunation is at hand. This usually occurs after the sun has set and the sky has darkened somewhat, but while the moon remains slightly above the western horizon. It appears as a thin crescent because, from our point of view, the moon is still mostly backlit by the more distant sun.

(I was walking with a five-year-old girl when she pointed in the sky to the moon and said "croissant." Clearly she was used to going to better restaurants than I was.)

The young crescent moon sets quickly, but on successive nights it is higher in altitude at sunset. Moreover, the crescent widens as the angle of illumination changes. Still, less than 50 percent of the lunar hemisphere facing us is illuminated.

Eventually, about a week after new moon, half of the earth-facing lunar hemisphere is lit and half is not. The lunar terminator is a straight north-south chord bisecting the moon. We call this phase . . . first quarter.

Quarter? Who said anything about quarter? Clearly, half of the moon's disk is visible. Yet that is not where the name first quarter moon comes from: At first quarter, there is a ninety-degree angle between the line intersecting the centers of the earth and moon, and the line intersecting the centers of the moon and sun—a geometry called a dichotomy. This happens when the moon is one-quarter of the way around its orbit. (Such vocabulary leads to confusion, with people using the term "half moon" when they mean quarter moon.) At sunset, the first quarter moon is near our celestial meridian.

Sometime in the third century BCE, a Greek geometer named Aristarchus of Samos had a clever idea: The earth, moon, and sun form a great triangle in space. At either dichotomy, it is a right triangle, meaning that

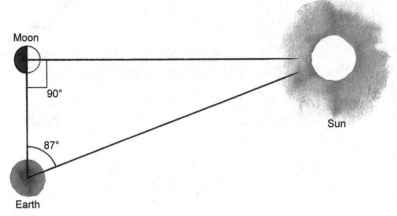

FIGURE 8.4. The geometry of Aristarchus's sun distance determination. Not to scale.

the side joining the earth and moon, and the side joining the moon and sun, are at a ninety-degree angle with respect to each other. Aristarchus understood that all right triangles behave in the same way. Their sides are proportional in a ratio fixed by the other angles in the triangle. If Aristarchus knew one of those angles, he would know the ratio, too. (He only needed to know one of the remaining two angles because the sum of the angles in all triangles is 180 degrees.) Aristarchus attempted to measure the angle between the earth-moon side and the earth-sun side. Doing so, he was poised to determine, for the first time, how much farther away from the earth the sun is than is the moon.

As it turned out, Aristarchus was a better mathematician than he was a surveyor. His measurement of the angle was much too large. (It is just a fraction of a degree—hard to measure at all by eye.) Moreover, he did not realize just how difficult it is to estimate when the lunar terminator is completely straight and the moment of dichotomy has arrived. Aristarchus's answer was wrong; he concluded that the sun is only eighteen to twenty times farther from the earth than the earth is from the moon. That is much too small a number. However, his ingenious method for making this determination was, theoretically at least, completely sound.

Our lunation continues past first quarter. Now, more than 50 percent of the lunar hemisphere facing the earth is illuminated by the sun. The appearance is a cookie-cutter reversal of the crescent moon—the bright lunar disk looks as if a crescent-shaped piece has been removed.

The name of this phase is known to fewer people than the number who are familiar with the name of the crescent moon. That is somewhat surprising because this phase is visible just as often as the crescent. And it can appear high in the sky, unlike the crescent moon, which is best

FIGURE 8.5. Proof that gibbous moons can be pretty, too. Taken in Derbyshire, England. Photo by Michael G. Lancaster.

seen hugging the horizon. Perhaps it is because of the word's strange sound: When the moon is more than 90 degrees (but not 180 degrees) along its orbital path, we call it the gibbous moon.

As further evidence of discrimination against the gibbous moon, I only have to point to the movies. In film, the moon usually is depicted as a pretty crescent. Sometimes it is full—if the mood is supposed to be spooky. Alas, how often have you seen a gibbous moon portrayed in a movie? In any visual art? Artists feel that the poor gibbous moon just does not look right.

The gibbous moon appears in our eastern sky before sunset. It is, of course, wider and brighter than the crescent moon. This width and brightness increases from night to night, as more and more of the sunlit moon becomes visible to us earth dwellers. Somewhat compensating for the gibbous moon's prominence in the sky is the fact that it appears, at sunset, closer and closer to the eastern horizon.

Next comes the full moon. It is our second syzygy of the month. The moon is halfway through its circular path around the celestial sphere; so, perhaps this phase should be called "second quarter." However, the fact that (a) the entire lunar hemisphere facing us is now illuminated and the moon is at its brightest, (b) the moon is a complete disk in the sky, and

FIGURE 8.6. An airliner coincidentally passes in front of the full moon. Taken in Sonnenbühl-Genkingen, Germany. Photo by Martin Wagner (www.himmel-und-erde.com).

(c) the moon at this phase rises as the sun sets, putting it above the horizon for the whole night, makes the name full moon an obvious choice for the phenomenon. Notice that while one crescent or gibbous moon can look different from another (in width, for instance), full moons (and quarter moons) are unique events.

(One disclaimer: Because of atmospheric refraction, it is just possible to see the full moon appear to rise before the sun has appeared to set totally. Refraction raises the image of both the sun and moon; so, near the horizon, the two luminaries look a little closer than 180 degrees apart.)

Technically, there is a single moment when we can see the maximum amount of lit moon, just as there was (technically) a single moment when the moon's face was exactly half illuminated (or, alternatively, the lunar terminator was perfectly straight) at first quarter. While the crescent or gibbous phases last for days, the full moon and quarter moons are instantaneous. Practically speaking, though, there will be one whole night, or even a couple of nights, during which the moon looks full (or approximately quarter) to the unaided eye. This is the more common definition of full moon.

How full is full? The moon orbits the earth in a plane different from the sun-earth plane (the ecliptic). So even when it is 180 degrees away from the sun in its orbit, the moon may be a little above or a little below the sun-earth line, causing a defect of illumination at the top or the bottom of the full moon. Nevertheless, I challenge anyone to detect this deviation from a perfect, bright circle at full moon.

No, people do not become moonstruck at full moon. The idea is a long-lived one, as can be seen in the etymology of "lunacy," from the Roman (Latin) word for the moon, *Luna*. Objective statistical studies of things like hospital emergency-department admissions show no correlation with moon phase. Sadly, people do crazy things to themselves and others every night of the month.

This is not to say that moonlight has not affected human history. Its presence or absence has helped or hindered, depending upon who was doing what to whom. Astronomer Donald W. Olson studies such things. He discusses how the nearly full moon's light facilitated the successful Japanese attacks on Pearl Harbor and later on the *USS Indianapolis* in World War II. Or how the just-past full moon illuminated American Revolutionary Paul Revere's legendary late-night ride. (On the other hand, a dim moon did not illuminate the goings-on at the Boston Tea Party.) Olson has further used moon phases to date and time famous artwork, such as Ansel Adams's photograph, *Moon at Half Dome*.

Our lunation continues. Now at sunset, the moon has not yet risen. For a time, there is no moon in the sky. When the moon does rise, it is in the gibbous phase again. The geometries that created the phases during the first half of the month now repeat, in reverse order, during the second half of the month. On successive nights, the gibbous moon grows thinner and appears later.

The moon is eventually 270 degrees from its starting point at the beginning of the month, a second dichotomy. The third week, third quarter moon (alternately, last quarter moon) looks like the first quarter moon, though it appears on the celestial meridian at sunrise, not sunset. Nevertheless, there is a difference: As the moon goes from gibbous to third quarter to crescent, notice that it is the opposite side of the moon that is illuminated compared to when it changed from crescent to first quarter to gibbous.

In Mesoamerican Aztec imagery there are nine characters named the Lords of the Night. The icons that stand for them look as if they may have something to do with the moon. One interpretation of the Lords of the Night is that they represent the evenings on which there is no moon.[5] Beginning at third quarter, the moon rises at midnight or later. From third quarter to new moon (about seven days), and then to the beginning of the new lunation (about two more days), most people will go to sleep at a dark, moonless hour. There will be about nine such nights per month, the sum of seven plus two.

It is the crescent moon again. We see it hanging low over the eastern horizon before sunrise brightens the sky and causes the moon to disappear. (If you do see the moon in the daytime, it is likely to be a respectful angular distance away from the sun—somewhere between first quarter and third quarter.) The cusps (horns) of the crescent moon always point away from the sun; in this crescent, the cusps are leading the moon instead of trailing.

Geometrically, the outer curve of the crescent moon always should extend 180 degrees around the circumference of the lunar disk. However, French astronomer André-Louis Danjon (1890–1967) documented a phenomenon that likely had been observed casually long before: thin crescent moons that extend, cusp to cusp, less than 180 degrees. In other words, a line connecting the points of the two cusps does not always pass through the center of the lunar disk. The effect is more pronounced the younger the moon is. Experts debate the cause of this phenomenon, but agree that, because of it, there is a theoretical angular distance from the sun at which no crescent can be observed. This angular distance is called the Danjon limit and is about seven degrees.

Insofar as the lunation begins and ends with a crescent, it is not odd that two-horned beasts are associated with the moon. In Paleolithic rock art, the horns on animals often are painted to look a lot like crescent moons. The Roman god Luna frequently was portrayed riding a chariot pulled by cows; and, as planetarium director Ed Krupp reminds us, in the old nursery rhyme, "The Cow Jumped Over the Moon."[6]

Our lunation has ended. Eventually, the moon is new once more. There may be as many as four nights (centered on new moon) between the last visibility of the crescent moon in one lunation and the first visibility of the crescent moon in the next lunation. However, if you have a good eye, you may have to wait as little as a day and a half between appearances of crescents. If the instant of new moon happens to occur at midday, it is sometimes just possible to catch sight of the crescent moon in the morning, and then again in the evening of the next day—with the new moon occurring in-between. The opportunities for such back-to-back crescent moons are rare; they occur on average only once in fifteen years. Even then, sky-observing conditions must be ideal.[7]

How do you tell one lunar phase from another? Nobody will confuse a new moon with a full moon. However, what about the first and third quarter moons? What about the pairs of crescent and gibbous moons?

It can be done with just a bit of practice. Where and when the moon is in the sky, and which part of the moon is illuminated, help. The most straightforward way, though, is this: Are you seeing less and less of the illuminated lunar hemisphere night after night? If so, it is the last part of the month, and the moon is said to be waning. Are you seeing more and more? The moon is said to be waxing. Note that this applies to both crescent and gibbous moons. To use their full names, the waning crescent moon is distinct from the waxing crescent moon; the waning gibbous moon is different from the waxing gibbous moon; first quarter is a waxing phase; third quarter is a waning phase. Since lifestyle causes more people to look at the evening sky than at the morning sky, we are more likely to see the waxing phases of the moon than the waning.

By the way, quarters are not the only way in which to divide the month. In Hindu, the half month from new, through the waxing phases, to full is given its own name—the *Suklapakṣa*. The half month from full, through the waning phases, to new, is the *Kṛṣṇapakṣa*.[8] The Batak (of Sumatra) are an example of a people who divide the month into ten-day thirds.[9]

Synodic months may seem anachronistic in a modernity slaved to the solar calendar. However, for hundreds of millions of people, the lunar calendar remains relevant. The most sacred holiday of Islam, Ramadan, occurs every twelve synodic months. It happens independently of the time of tropical year, and shifts backward through the seasons. For example, Muslims celebrated Ramadan twice in 1999: The month-long observance fell December 20, 1998, through January 18, 1999, *and* December 9, 1999, through January 7, 2000.

MYTHS OF THE MONTH

Are plants affected by the phase of the moon? No. Probably not. While almanacs traditionally provide planting instructions based upon moon phase, a connection is dubious.

Tides? Yes. The gravity of the moon and sun do create ocean tides. At syzygy, these forces combine to produce especially strong tides. The sun and moon are the only bodies both massive enough and close enough to the earth to produce any tidal effect, and it is the nearby moon that is responsible for most of it.

FIGURE 8.7. The moon in all its phases. Photo by António J. Cidadão.

Tides affect an entire ocean. Tides happen because of the difference in gravitational force between the moon and the ocean on the far side of the earth and that on the near side of the earth. They are a differential force. For tidal effects to apply to (say) the fluid within such a tiny length as a plant stem, the moon's gravity would have to be immense. It would certainly kill us—human beings with bodies composed principally of water.

On the other hand, let us imagine that we are told to plant tomatoes on the third full moon after the vernal equinox. That full moon might occur on a date between May 18 (for Northern Hemisphere tomatoes) and June 17. It may well be that planting on any of those days might result in county-fair-grade tomatoes. But it is the time of year, and not the synodic month, that is important in determining the time to plant.

Another adage about the moon tells us that we can forecast the weather by whether the cusps of the moon are pointed downward or upward. "Rainy down, fair skies up." The upturned cusps hold the water back? The moon cusps point in both directions equally often. So in reality the saying is telling us that there are equal chances of rain and sunshine. In some climates this may be true. However, there is no correlation between the moon's orientation and weather.

9

Living Month to Month

New moon on Monday
And a fire dance through the night
I stayed the cold day with a lonely satellite

DURAN DURAN,
"New Moon on Monday," 1983

The month is a handy unit of time: an order of magnitude longer than the day, but an order of magnitude shorter than the year.

MONTHLY LISTS

In the 1960s, a twenty-thousand-year-old (or more) worked bird-bone fragment was found in France. This age places it in Paleolithic times. Alexander Marshak saw the bone in a museum and thought that the markings on it might be purposeful (not just, say, the sharpening a knife). He was the first to interpret them as monthly day counts, instead of something arbitrary like a hunting score (akin to our modern notches on a gun belt). If that is the case, it would make this find the oldest evidence of sky watching that we have to date.

Obviously there was a pattern of scratches; a microscope reveals that sets were made by different tools at different times. Why, though, jump to the conclusion that they are the work of a person keeping track of the days in the month?

If you count the synodic month by watching the sky, you will not, of course, get 29.5 days. Instead, sometimes you will get twenty-nine, and other times thirty—depending upon the exact moment you first see the waxing crescent moon, or the exact night you decide that the moon is full. This is exactly what we do see in the bone tallies: groups of twenty-nine and thirty. Similar, but newer, notched bones have been found in Egypt and elsewhere in Europe.

Plus, there may be examples of month counting in the New World. The Las Bocas pendant is a large piece of Olmec jewelry excavated in Mexico. It dates from circa 1000 BCE. This ancient "bling bling" is made of carefully matched pyrite tiles. Water exposure has damaged its original brightness—the pendant once was highly polished.

The mosaic of tiles is made up of groups of fifteen or sixteen.

Marshak believes that the half-month was important to prehistoric peoples as well (though certainly if you accept both months and half-months, you increase the odds for coincidence). Still, the total number of tiles equals 354, and this is the number of days in a "year" of twelve synodic months:

$$12 \times 29.5 = 354$$

In Mexico, there is a wall panel of Native American rock art today named Presa de la Mula. On it are marked more than two hundred strokes. In the 1980s, Presa de la Mula was interpreted by William Breen Murray as a count of seven synodic months. Other such panels exist in Mexico and the United States. Nobody knows how old they are.

Who made all these calendars? For they do resemble our calendars: Sometimes little pictures along the count scratches seem to illustrate an event associated with a date, just as we might doodle in one of the boxes on our printed wall calendar. What do you think the carver at Presa de la Mula was timing? How do you think he dealt with the times that the moon could not be seen? Or was it a she, tracking the time to child-bearing?

A lunar calendar is a perfectly reasonable calendar for use by hunting communities who rely on the moon's light to illuminate their prey. However, agricultural and herding communities might find more utility in a solar calendar, because of the importance of seasons to their food supply. The trick comes when a civilization wishes to fit both calendars into a lunisolar calendar.

Twelve synodic months equal 354 days. This is close to, but short of, the (approximately) 365 days in the year. Each year it will be necessary to start counting the twelve-month cycle over again earlier in the annum. Eventually, you will be at the end of month twelve, and there will be more than 29.5 days left in the year. One can imagine things getting so far out of hand that the *first* month of your twelve-month lunar calendar occurs in the *middle* of the tropical year. The two are completely out of synch. So as to keep this from happening, it becomes necessary to occasionally intercalate—insert an extra thirteenth month into the calendar. If this is done wisely, the first month in your twelve-month cycle always will remain early in the year, and the last month will stay late in the year. For example, in the Jewish calendar, the twelfth month of Adar sometimes is followed, not by the first month of Nisan, but by a thirteenth month called Adar (again).

Yet, intercalation is based on a human decision, and it can be misused for political reasons. The Koran expressly forbids intercalation. This is why the Islamic calendar bears no relation to the civil, solar calendar, and holidays (such as the *Eid*) can occur at any time of year.

The Greeks codified intercalation. Meton of Athens (circa 432 BCE) established a pattern in which intercalation would occur every third, then third, then second, then third, then third, then third, and then second year—a pattern that is then repeated. Such a system results in an average year 365.26 days long. Not bad,—but who could remember it? Callipus of Cyzikus (circa 370–300 BCE) went so far as to suggest that, if four of these Metonic cycles were followed, and then one day of the last year was dropped, the result would be a better average year of 365.25 days. No, he was not kidding.

Julius Caesar finally had enough of intercalation. It was the Romans (45 BCE) who assigned the months fixed lengths so that twelve months would fit exactly into 365 days. (Leap years were inserted according to a formula so that the average year was 365.25 days long.) This was a big deal because, before Julius (really his astronomical advisor, Sosigenes of Alexandria), intercalation was a means of correcting an error. Leap years are a means of prohibiting an error from occurring in the future. In an anticipatory (proleptic) calendar like Sosigenes's, you fatalistically accept that your calendar always will be wrong, and take steps to minimize that error.[1]

The proleptic Julian calendar was born. Today's Gregorian calendar is closely based upon it. The result, though, was a calendar in which the months no longer had anything to do with the real synodic month based upon the cycle of moon phases. Alas, you cannot get something for nothing.

Going further, perhaps the strangest "month" of all is the classical Chinese Ch'i. It is not so much a month as a division exercise. The Han took a 365.25-day year and divided it by twelve. The result is twelve 30.4-day-long Ch'i.[2] Here, not only is the physical moon ignored, but so is the requirement of integer numbers of days.

The Gregorian Calendar

The Gregorian calendar reset the count so that the vernal equinox occurs on March 21, and fiddled with the leap-year rule slightly so that the slippage would not occur again.

For example, the year 2000 was noteworthy not only for being the end of a millennium; it was also a leap year! That is not by itself surprising. We learned in school that an extra day (February 29) is to be inserted into every year that is divisible by four. Yes, but according to the Gregorian calendar of 1582, based on the correct 365.24-day year, we must skip a leap year every hundred years except in years divisible by four hundred. The year 2000 was a leap year, but it also was divisible by four hundred, which made it special. So was the year 1600. Even Pope Gregory's calendar eventually will go wrong—we likely will have to drop a leap year in 4000.

(That is, if this Gregorian thing catches on: England, not wishing to rush into anything, did not adopt the Gregorian calendar until 1752. In Russia, it was 1918.)

While the civil calendar may be done with the moon, many work weeks in the United States still are influenced by the date of Easter. Easter is defined as a spring holiday (solar calendar); but by tradition, it is to be illuminated by a bright moon (lunar calendar). This caused the Christian church to come up with a rather complicated definition: Easter is the first Sunday following the first full moon after the vernal equinox. (The Orthodox Church may observe Easter on a different date than other Christian churches do because it still uses the Julian calendar.) Insofar as the holidays of other major religions also are regulated by the synodic month, the moon and our calendar are not divorced totally.

Our modern calendar received a final tweak during its separation from the real moon. To five places, the year is 365.24 days long. That difference of one hundredth of a day added up enough by the Renaissance so that the seasons were no longer matching their appropriate dates.

A BLUE MOON

This discussion of the so-called blue moon is much longer than the importance of the phenomenon would seem to dictate. Still, I get a surprisingly large number of questions about this particular term.

The moon is largely colorless. As it reflects yellowish-white sunlight, it looks yellowish-white to our eye most of the time.

Why, then, is the phrase "blue moon" so popular? It has certain alliteration—I grant you that. Also, it conjures up an appealing, lonely-yet-romantic feeling. Rodgers's and Hart's "Blue Moon" must have been recorded more times than any other song except "Happy Birthday." (My favorite version of "Blue Moon" is by the Marcels.)

But what does the phrase mean? In the sixteenth century, when it first appeared in written form, "blue moon" equated with an absurdity: "Yeah, right. And the moon's blue, too!" (Today we might say instead, "like frost on the Fourth of July" or "when hell freezes over.") "Absurd" evolved into "never."[3]

Yet, by the nineteenth century, English-speaking people traveled more. They realized that the moon occasionally does turn blue. (More on this later.) So, the definition of "blue moon" changed to that of a rare, and not entirely predictable, event.

As for the blue moon's modern association with the calendar, it was so used on the public radio program *StarDate* in 1980. It appeared in a popular children's book in 1985, and received the honor of an appearance on a Trivial Pursuit game card in 1986. By about 1988, many people thought that the blue moon had something to do with the calendar—even if they did not know exactly what. The Blue Moon Café, on my own street in Cedar Falls, Iowa, was so named around that time.

What is the association with the calendar? The *Maine Farmers' Almanac* used "blue moon" in the 1930s. (We do not know from where the MFA got it.) Still, that reference meant something different from the calendrical term used today.

According to the *Almanac*, the third full moon in a season with four full moons is the blue moon. The full moons in a given season have traditional names. There is the harvest moon in the fall, worm moon in early spring, egg moon near Easter, and so on. Such a system assumes twelve full moons per year. If we do not call the extra full moon something else, the pattern gets out of "whack." For example, the full moon that is supposed to be the hunter's moon might occur too early to illuminate the seasonal migration of animals.

Today's slightly different definition, the second full moon in any calendar month, is based on a thirty-year-old mistake. The author of an article in a popular astronomy magazine, describing the blue-moon concept, apparently misunderstood the rule that the almanac writer had been using. But it is the second-full-moon-in-a-month rule that stuck.

Why do second full moons happen at all? Because we use calendar months of thirty or thirty-one days, the real full moon will occur earlier and earlier in the calendar month. If a full moon happens on, say, June 1, the next full moon will do so 29.5 days later, on June 30. It is still June.

Blue moons defined this way are not all that rare. In a thirty-one-day month, it is even easier to have a blue moon. The first full moon could occur on the first or even the second day of the month. The only month in which a blue moon is impossible is February. On the average, there is a blue moon every two or three years.

Totally unrelated to the calendar is the fact that the moon literally can turn blue. This sounds bizarre because we are used to the moon, if it appears to have any color at all, being yellow, orange, or red—on the opposite end of the spectrum from blue. Normally, the particles in the earth's atmosphere scatter blue light. All the blue light reflecting from the moon scatters out of our line of sight, leaving just the redder light to pass straight from the moon to us. But occasionally, the particles in the air are just the right size to do it the other way around: Red light is scattered, and blue light passes directly through from the moon to us. Examples are the particles inserted into the air by volcanoes or forest fires. Reports of truly blue moons have been associated with both of these— hopefully—rare events.

Incidentally, sunlight, too, can be turned blue under extreme conditions—such as a sandstorm; however, I have never seen it do so. "Once in a blue sun"?

Blue moons are not astronomically important. I searched a database of professional papers going back a hundred years, and in it, nowhere does the phrase "blue moon" appear. Still, I think that the existence and use of the phrase tells us something about popular culture.

WHENCE THE MOON?

Moonrise and moonset have a lot in common with sunrise and sunset. When I am called to trial as an expert witness, the dispute is usually an evening automobile accident. The question of interest normally has been, "How bright was the moon?" In other words, was there enough light by which to see? (This scenario happens less frequently now than it

FIGURE 9.1. Are we meant to infer that this beer brand is of rare quality? Photo by Michael Hockey.

did in the past—such information is readily available today on the World Wide Web.)

As the moon's distance from the earth varies little, the two significant factors that influence moon illumination are phase and moonrise (or moonset) time. Each night the moon rises and sets later than it did the night before, but the difference in time varies. The moon travels on its orbital path at a nearly constant rate. However, its motion with respect to the horizon determines the moonrise or moonset delay in hours or minutes.

The ecliptic can be nearly perpendicular to the horizon or oblique to it, depending on latitude and time of year. The moon's path with respect to the horizon varies over an even greater range of azimuth than does the sun's because the moon is not on the ecliptic, but somewhere within five degrees of it. The component of the moon's daily motion perpendicular to the horizon is what determines how much later moonrise and moonset will occur night to night.

Presently, during northern autumn, the ecliptic and moon's orbital path conspire such that little of our satellite's daily orbital motion is perpendicular to the horizon. Most of it is parallel. So at a given hour, from night to night, the moon has not changed its position much with respect to the horizon. Moonrise on one night is only slightly later than it was on the night before. We tend to notice this more—as well as everything else that has to do with the moon—at full moon.

Near the autumnal equinox, there will be two, or even three, nights when a nearly full moon is rising approximately at sunset. This is the famous harvest moon—the full moon closest to the equinox. There is some ambiguity regarding the traditional timing of the harvest moon: Is it the first full moon *after* the equinox—or the one *nearest* the equinox (before or after)? Nevertheless, it provides several nights of bright moonlight to help farmers harvest their crops—if the farm is at a latitude such that harvest occurs during this time of year.

There is reason why we northerners associate the full moon more with Halloween jack-o'-lanterns than we do with, say, Easter eggs: We see it more often in the fall. In the spring, much of the moon's orbital motion is perpendicular to the horizon. The moon is in a very different place with respect to the horizon night after night. While we may well see a March full moon, if we look for it at approximately the same hour the next night, it will not yet have risen. The times between successive moonrises (and moonsets) are greater near the vernal equinox than near the autumnal equinox.

The two variables of the ecliptic's orientation with respect to the horizon and the orientation of the moon's orbital path with respect to the ecliptic also determine at what altitude the moon will culminate on a given evening. Full moon is the easiest example. At full moon, the moon is opposite the sun with respect to the earth. In the temperate winter, the sun culminates close to the horizon. In the Northern Hemisphere this is because the sun is well south of the celestial equator at that time of year. Yet the ecliptic encircles the earth; if half of it is above the celestial equator, the other half must be below it. In the winter, this part of the ecliptic does not house the sun, but at full moon it is the approximate address for the moon. So twelve hours after noon, the full moon, always within five degrees of the ecliptic, must be *high* in the sky. Everything is reversed in the summer.

Do not just take my word for it. Most northern temperate dwellers can picture that one perfect January night: It is clear and cold—the kind of night when you can hear your footsteps crunch against the snow. And all the snow crystals glisten, because of a full moon high over head.

Now fast forward to a summer's evening. Where is the full moon? It is slinking along close to the horizon. If it is hazy (not a long shot in summer), the full moon is dimmed. Near the horizon, the odds of its being obscured by clouds are increased. There are as many full moons in the summer as there are in the winter, but we are less likely to notice them.

(Admittedly, in my scenario, the shiny winter's snow helped bring attention to the moon, too.)

I have mentioned tropical peoples to whom dates on which the sun is at the nadir are important. You cannot see the sun at the nadir. However, the full moon may act as a stand-in: If the full moon is seen at the zenith, the sun must be near or at the nadir.

MOON RISE MADE EASY

The moon's circular apparent path against the shell of stars is tilted such that half of it is closer to the north celestial pole and half is closer to the south celestial pole (similar to the ecliptic). On those days (half of the sidereal month), when the moon is on that part of its path closer to the north celestial pole, it will rise in the northeast. On those days (the other half of the sidereal month) when the moon is on that part of its path closer to the south celestial pole, it will rise in the southeast.

When the moon is as far north as it will travel during the sidereal month, we say that it is at its northernmost standstill. When the moon is as far south as it will travel during the sidereal month, we say that it is at its southernmost standstill. (The term "lunastice" never caught on.)

At a lunar standstill, the moon appears to change directions. Night after night it rises more and more southerly. Then it reaches its standstill. It will, thereafter, rise more and more northerly. The moon does this until it reaches a second standstill, at which time it will begin to rise more and more southerly again. Think of the direction of moonrise as another swinging pendulum. The cycle repeats every sidereal month, with the two standstills (northern and southern) happening about two weeks apart. Note that everything I have said about moonrise also applies to moonset.

So far, the moon seems to mimic the behavior of the sun insofar as the sunrise and sunset directions also swing back and forth across segments of the eastern and western horizons. (Of course, the standstills for the sun go by the name of "solstices.")

Now here is where the behavior of the moon becomes more complicated than that of the sun: The entire orbit of the moon twists around the earth in a big circle every 18.61 years. (This is a gravitational effect, not unlike precession.) So sometimes the apparent path of the moon in the sky is closer to the celestial equator than it is at other times. When it is closest to the celestial equator, the southernmost and northernmost lunar standstills do not differ much. They huddle close to the celestial equator. When the position of moonrise on the horizon varies the least, at such a time we call the two standstills (now closer to the celestial equator than they will be at any other time during the 18.61-year cycle) the minimum southern standstill and the minimum northern standstill. When the apparent celestial path of the moon is farthest from the celestial equator, though, the southernmost and northernmost lunar standstills differ greatly from one another. The moon rises far to the southeast one night and, half a month later, it rises far to the northeast. When the position on the horizon of moonrise/moonset varies the most, at such a time we call the two standstills (now farther from the celestial equator than they will be at any other time during the 18.61-year cycle) the maximum southern standstill and the maximum northern standstill.

Every 9.3 (18.6/2) years, there will be a month in which the moon arrives at either a minimum or maximum lunar standstill. But wait. Half of those moonrises or moonsets will occur in the daytime. Odds are that some will happen during cloudy weather. Regardless, the scene will not repeat until the same window of opportunity opens 18.61 years later. Thus, catching the moon at a standstill is tough. The only thing making it easier is the slowing of the lunar pendulum near standstill, so that there are, in practice, several nights in a row during which the moon is near its standstill rising or setting azimuth.

Note that the inclination of the moon's orbital path in the sky to the celestial equator is also what determines at what altitude the moon crosses the celestial meridian. The maximum altitude of the moon on a given night is also governed by the 18.61-year cycle.

One can imagine being so far north or south that, at times, the moon does not rise above the horizon at all during the course of one or more days. This latitude corresponds to the Arctic and Antarctic Circles in the case of the sun. However, with the moon, the "no show" circle of latitude varies, reaching extremes at the maximum and minimum standstills. Notice that it then must also be possible for there to be times at these latitudes when the moon does not set over the course of one or more days.

The difference between the solar circles and their lunar counterparts is that the Arctic and Antarctic Circles remain fixed on the globe for a long time. The lunar standstill parallels vary from month to month. We need yet another circle of latitude, this time a lunar standstill maximum (or minimum) circle, which remains constant. However, the globe may be crowded enough as it is.

WHO IS MINDING THE MOON?

Looking for architectural alignments to the moon is fraught with peril. Whatever the range of azimuths at which the sun may rise or set from

FIGURE 9.2. Chimney Rock, Colorado. The northern lunar standstill moonrise as seen from the Ancestral Pueblo Great House ruin. Photo by Helen L. Richardson.

a given location, the range at which the moon may rise or set is greater. The potential for coincidental orientations increase.

Remains of an Ancestral Pueblo community in Colorado are situated within sight of a two-spired butte now called Chimney Rock. There, at lunar standstill maximum, the moon rises spectacularly between the two tower-like formations. The apparent gap between the chimneys is wider than the apparent size of the moon. So the lunar standstill will get far enough northward about fifteen months before the maximum such that the moon will fit within the chimneys at moonrise (on one or more days during that month). And it will remain far enough northward about fifteen months after the maximum to fit within the chimneys at moonrise (on one or more days during that month). (The fact that the bottom of the space between the chimneys is not exactly at the astronomical horizon has a small effect on the phenomenon.) The orientation duration benefits from the slowing of the rising moon's pendulum-like swing along the horizon at either standstill. Even so, there are only thirty northernmost standstills per cycle during which we can see this phenomenon (if visible).

The prehistoric inhabitants near Chimney Rock may have noticed the phenomenon. They may even have celebrated it. Nonetheless it is another thing entirely to suggest that they purposely situated their settlement where they did, expressly so as to align with Chimney Rock and the lunar standstill. This is a stretch insofar as they might then just as well have homesteaded a spot where the moonrise was visible through the chimneys frequently.

There is no topic more controversial among archaeoastronomers than the lunar standstills. Did ancient people actually pay attention to them? Did they go to the trouble of recording standstill maximum or minimum azimuths in monuments that remain as evidence for us to see?

It would take a while to do so—at least nineteen years. That is not a trivial thing, considering the short average life spans during the eras in question. Even then, the standstill is hard to pick out, unless the moon is full (and the weather is clear) at the right time. Unlike sunrise, where dawn provides some warning, moonrise appears suddenly at the horizon. It can sneak up on you. Moreover, what do you measure? Is it the center of the lunar disk? Again, this is difficult to do unless the moon is always full. Is it the "edge"? Same problem: It is hard to decide where the appropriate "edge" is, in different phases.

Despite all of this, Alexander Thom found all sorts of stone circle orientations to the moon in Britain. Of course, if both lunar and solar exist, you might expect to find more lunar lineups than solar ones. While there are only two solstices, there are four interesting lunar standstill azimuths, corresponding to standstill-maximum moonrise, standstill-maximum moonset, standstill-minimum moonrise, and standstill-minimum moonset.

Notable among Thom's lunar sites is Callanish, in the Scottish Hebrides Islands. At this one-of-a-kind place, there are some compelling orientations, including those to the equinoxes. However, the most elaborate seem to concern the moon: A set of parallel upright stones, their faces radiating out from the center of the monument, point toward the southern standstill maximum. Some have called these paired menhirs an "avenue."

It was Gerald Hawkins (1928–2003), the astronomer famous for the popular book *Stonehenge Decoded* (1965), who drew attention to Callanish (and the moon) in an effort to argue that Stonehenge is not unique. The southern maximum standstill is especially significant at Callanish—more so than at any other proposed lunar site including Stonehenge. This is because Callanish is so far north that, at maximum southern standstill, the moon is far south of the equinoctial azimuth. Indeed, it is so far south that it just barely appears—briefly, at low altitude—above the horizon. Then it sets with little difference in azimuth between moonrise and moonset. But at Callanish, the natural horizon is craggy; when it first sets, the moon does so behind a mountain. The act of setting also means a change in azimuth. The moon then reappears, briefly, within a gap in the mountains,

The Slivery
Moon

Especially because of its significance to the Moslem calendar, it is useful to be able to predict on which evening the first sighting of a waxing crescent moon will occur. For instance, the ninth Islamic month, *Ramadan*, is a month of fasting. Without a means of predicting ahead of time, the faithful do not know when to commence or break the fast until the waxing crescent moon is spotted (by two trustworthy individuals, tradition states) only the night before.* Different people at different places easily can end up observing the holiday on a different set of days.

However, prediction is harder than it sounds. Lunar visibility is a function of how much light the moon reflects our way, but whether we can make out that light also is governed by things that control contrast, such as twilight and extinction. Simply waiting a number of hours past new moon—twenty-four is a traditional value—does not work. A ten-hour-old moon on the ecliptic has almost the same brightness as a (nearly) zero-hour-old moon five degrees away from

the ecliptic. The angle between the sun and moon is the important variable; but even at a fixed sun-moon angle, the width of the lunar crescent varies, depending upon how close the moon is to the earth. (That distance does change; see the next chapter.) The interval between sunset and moonset is equally undependable. In fact, there does not appear to be one single parameter that allows prediction of crescent-moon visibility: Some combination of age, lag-time between sunset and moonset (or sunrise and moonrise), lunar altitude, sun-moon angle, difference in altitude between the sun and the moon, difference in azimuth between the moon and the sun, and crescent width is required.†

*N. Guessoum and K. Meziane, "Visibility of the Thin Lunar Crescent: The Sociology of an Astronomical Problem (A Case Study)," *Journal of Astronomical History and Heritage* 4, no. 1 (2001): 1.

†Mohammad Odeh, "New Criterion for Lunar Crescent Visibility," *Experimental Astronomy* 18 (2004): 39.

FIGURE 9.3. A lunar occultation. The planet Venus apparently disappears behind the moon. It is, in fact, the moon doing most of the moving from our point of view. Photo by Dave Weixelman.

before it sets for good beneath the astronomical horizon. This unexpected gleam (reappearance) of a bright moon is conspicuous. During the gleam, the moon is framed by the Callanish "avenue." At Callanish, latitude and the local horizon conspire to produce a one-of-a-kind effect.

At the end of the day (or should I say, "month"?), what needs to be answered is this awkward question: Why, exactly, should our ancestors have cared so much about standstills? One possible explanation is for the purpose of attempting to predict eclipses. However, that is quite a further leap in computational ability: Even today, successfully predicting an eclipse takes some serious mathematics. Did our ancestors *need* a practical reason? The Callanish moonrise effect would not be the same even a short distance from Callanish. Would the Callanish builders have built "a Callanish" anyplace else? The stones are silent on these matters.

MOON OVER . . .

While it is dangerous to stare at the sun, one can safely watch the moon (closer than all the planets) move in front of a planet or star and make it, temporarily, disappear. Such an event is called an occultation. A lunar occultation is more easily observed at immersion, when it is the dark limb of the moon passing over the planet or star, rather than the bright (and glaring) crescent or gibbous limb. Similarly, it is more easily observed at emersion, when it is the dark limb of the moon uncovering the planet or star, rather than the bright (and glaring) crescent or gibbous limb. During a grazing occultation, when the lunar limb brushes a star tangentially, the star may even flicker, as it alternately disappears behind moon mountains, and reappears within moon valleys, in profile on the edge of the lunar disk.

A near occultation—so near that our eye cannot separate the two bodies—is called an appulse. Historical records of lunar appulses with stars may be of use to astronomers even today. By knowing the exact location of the moon on the celestial sphere long ago, they can detect tiny deviations in the moon's orbit.

The Turkish (as well as Malaysian) flag incorporates the Islamic symbol of the star and crescent. This sign seems to predate the Ottoman Empire. Some have interpreted the symbol as representing a historical appulse of the planet Venus and crescent moon. However, what is wrong with such a literal interpretation is that such appulses are not uncommon. There is much disagreement over which historic date is the significant one.

10

Facing Up to the Moon (and the Sun, Too)

Even the Man in the Moon disappeared somewhere in the stratosphere.

SHINETOWN, "Second Chance," 2008

Unlike the stars, and the planets, the moon is a bright extended object even to the naked eye. That is, we can see that it has an apparent angular extent. The moon has a face. To a less noticeable degree, so does the much brighter sun. This distinction between the two bodies, disklike as opposed to the other "planets" (objects moving on the celestial sphere), earned them long ago a combined name: Considered together, the sun and moon are the luminaries in the sky.

BIG AS THE MOON

Particularly at full moon, we are confronted with a phenomenon that causes us to question what we see with our own eyes. The rising full moon is beautiful. Because of the convenient times of day at which they occur, we frequently see full moon risings and settings, and we comment on how big the moon looks—perhaps twice as big as it appears when it is far from the horizon.

We are wrong. The moon does not appear appreciably larger when rising or setting. In art, the full moon is routinely portrayed as overly large, compared to other astronomical objects. This may be done for truly artistic reasons. But, here, by the "moon illusion" I mean a real trick of perception. Indeed, it is absent in photographs.

It *may* be possible to judge the change in the moon's apparent size due to the eccentricity of its elliptical orbit. The moon is closest to the earth at perigee and farthest from the earth two weeks later, at apogee.

However, the eccentricity of the moon's orbit is slight, and observation of the moon's varying apparent size escapes all but the most dedicated moon watchers. (Astronomer Kevin Krisciunas points out that a cross-staff with a hole punched through the moveable bar would have made measuring the changing apparent size of the moon much easier—but nobody in antiquity appears to have done this.[1])

Besides, the moon's apogee and perigee would not explain why the moon always looks bigger near the horizon. In truth it should look smaller: The moon is slightly farther from us then—by one earth radius—compared to when the moon is at the zenith. This one-earth-radius difference in distance (2 percent) has an imperceptible effect on the moon's apparent size.

No, the big full moon is an illusion perpetrated by our brains. It can apply to the sun and compact groups of stars like the Pleiades near the horizon, too. (Once upon a time it was called the "celestial illusion.") What do these things have in common? They are far enough away that our binocular vision, by which we usually judge distance, fails.[2]

Does the moon illusion have to do with color? Just as the air scatters the blue light of the setting sun, allowing its red light to pass through, the same thing happens to the moon. The moon is not bright enough that its blue light turns the sky blue, but its disk definitely can look red.

So do we see redder objects as bigger? Not necessarily. The moon is intrinsically pretty gray. High in the sky, the full moon is whitish-yellow (just like the sun). If we look at this full moon through red-tinted glasses, it does not seem to suddenly grow in size.

What is the deal with the horizon? Is the moon illusion due to atmospheric refraction? This effect may make the moon look out-of-round (oval), but it does not enlarge it.

The moon illusion is not fully understood. Yet a key to it is the fact that it is most commonly observed, not just when the moon is on the horizon, but when the horizon line is not quite dark (dusk or dawn)—that is, when nearer, terrestrial objects also can be seen. Apparently we think the moon is bigger when there are other familiar objects in the foreground: houses, trees, and smokestacks on the horizon. High in the sky, there are no such familiar objects nearby to mislead the eye. We see the full moon, as the apparent size it truly is, at midnight—not on the horizon.

Not convinced? Try looking at the rising or setting full moon upside down. This can be accomplished by turning your back on the moon, bending down, and looking between your legs. A "huge" moon magically shrinks. You have disrupted the view our brain expects to see—a hori-

zon and horizon features below the moon, and it fails to produce the illusion.

This test should be used with caution. I once demonstrated it to
a friend as we watched the moon rise, through his picture window. He
joined me in "mooning" the moon. And his mother-in-law was faced with
a less-than-dignified scene upon walking into the room!

MOONLIGHT AND MOON MARKS

The moon surely is shiny. And its brightness is predictable—you can
count on it. While the moon varies in distance from perigee to apogee as
it revolves around the earth, this difference is not enough to affect appreciably how much light the moon shares with us. The full moon is always
nearly the same brightness (assuming that it is not diminished by our atmosphere), the first quarter moon is always about the same brightness,
and so on.

(We all have been told that the rainbow is produced by the light
of the bright sun. I have seen nights on which the moon was so bright,
contrasted with its surroundings, that—in the clearing following a rain
shower—it produced a "moonbow.")

Yes, you can see at night by full moon light. However, compared to
the sun, our satellite is fairly dark: The moon only reflects back 7 percent
of the sunlight that reaches it. Quantitatively, astronomers say that the
average albedo of the moon is 0.12. It is a pretty bad reflector.

The average albedo of the moon changes with phase. At full moon,
it is high noon in the middle of the lunar disk. However, the surface of
the moon is rough; once the angle of illumination is no longer vertical,
all those little shadows, produced by lunar relief features, add up. This
shadowing and the obliquity of illumination are what cause the overall
brightness of the moon to decrease markedly as one gets further away in
time, waxing or waning, from full moon. The quarter moon is only about
one ninth as bright as the full moon.

Even at a given phase, the albedo of the lunar disk is not the same
everywhere on it. The surface of the moon is divided by albedo: slightly
lighter and slightly darker. (This is most noticeable at full moon.) Interestingly, third quarter is not as bright as first quarter. The hemisphere of
the moon that is sunlit during third quarter is intrinsically darker than is
the hemisphere lit during first quarter.

The unaided eye easily picks out the major albedo variations on the
moon, though ancient cosmologists who thought that celestial bodies

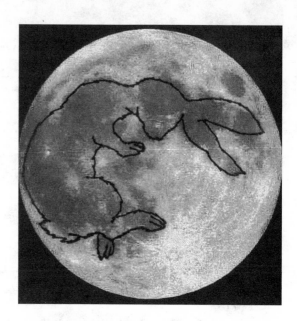

FIGURE 10.1. One way to envision the Rabbit in the Moon.

must be perfect (unmarred) sometimes tried to ignore this. Still, most children can play the game of identifying the Man in the Moon (or seated lady or rabbit—depending on your culture) made up of the light and dark albedo regions on the lunar disk.

The position angle of the moon refers to the orientation of its disk. It is the angle true north on the moon makes with respect to our horizon. The moon's position angle changes as it appears to move across the sky. (To see a different position angle, look at figure 8.6.) This influences what we think we see in and among the high and low albedo regions. The most common morphologies recorded are those visible at rising or setting position angles.

With no atmosphere to protect it, the moon is constantly pelted by the impact of small solar-system bodies. The most heavily-cratered, highest-albedo terrain is called the lunar highlands. (Its mean elevation is a little greater than that of the other type of terrain.) The other, lower albedo terrain is called the lunar mare. Maria are smoother than the highlands and vaguely round.

Maria are thought to have been formed by molten rock: Dark lava welled up from inside the moon after the mare basin was excavated by a giant impact. Planetary scientists believe that the maria are more recent than the highlands based on the following argument:

The moon is covered with craters, though only a couple of big examples can be resolved by a good eye. With no atmosphere to protect it, small solar-system bodies routinely strike the moon. These high-speed impacts blast out the craters. (The resolution of our eye is insufficient to make out individually most of these ubiquitous features.) Let us assume that the objects that create craters have done so randomly over the lunar surface throughout its history. Then, one day, a rare, much larger object hits, obliterating everything beneath. This surface is "repaved." The cratering continues. It still is random. However, for a long time into the future, we can tell which terrain is younger by the crater density.

This is an important method for scientists trying to date surfaces that we cannot visit. One also can date city streets this way. A heavily potholed pavement is almost always older than one that is not potholed, right?

About two-fifths of the moon we see from the earth is mare. It is the maria that create the silhouette figures we pretend to see on the lunar disk with the naked eye. For instance, Mare Serenitatis is the Man in the Moon's left eye, and Mare Imbrium is his right eye. These figures are harder to see during the glaring contrast caused by a full moon on a totally black sky. They are best picked out in a twilight sky.

"Maria" literally means "seas." These features were so named because early astronomers thought that the smooth, darker places on the moon were indeed oceans. They could not have been further from the truth—the moon is exceedingly dry. Astronomers did not realize that, when viewed from high up, the oceans are rather shiny.

Artistic representations of the moon date back to paleolithic rock art. However, these figures show the phase of the moon only. Amazingly, no Western artist painted or drew the moon as it authentically looks— blotches and all—until the Renaissance. (Some pre-Columbian art arguably shows the lunar rabbit.) It was long thought that the celebrated Leonardo da Vinci (1452–1519) was the first to do so. His famous notebooks depict the moon as the master saw it in 1513 or 1514. Art historian Scott L. Montgomery points out, though, that the Dutch painter Jan van Eyck (1385?–1441) portrayed a realistic moon more than a generation before Leonardo. The oil painting is *The Crucifixion* from circa 1420. It includes a gibbous moon, small in the painting, but with lunar mare of correct shape and position. Realistic moons can be seen in Van Eyck's later works, as well. All go beyond Leonardo's rough sketches.

It was not until around 1600 that English physician William Gilbert (1544–1603) created a map of the moon. That is, Gilbert drew a stylized moon but also added a cartographical grid and feature names. His names failed to stick. In 1609 Galileo Galilei (1564–1642) turned a telescope on the moon, and selenography (the study of the moon's appearance) once and for all transcended the naked eye.

Did the
Moon Blink?

With no geologic processes currently at work there, the face of the moon now is essentially changeless—or so we think.

June 1178: A group of monks in Canterbury, England, witnessed something extraordinary. In the words of Gervase (circa 1141–1210), their chronicler, the "upper horn of a new moon split and from the division point fire, hot coals, and sparks spewed out."

Geologist Jack Hartung interprets this story as evidence for a huge, explosive impact on the moon. He used the timing of the event plus data on the moon's orbit to estimate its geographical location on the lunar globe, and to link it with the presence today of a fresh twenty-kilometer-diameter crater at that spot.

Others have pointed out that, statistically, the formation of a large impact crater caught in the act anywhere on the moon, within the last millennium or so, would be a fantastic coincidence. (What may be common over spans of geologic time can be extremely rare over historical time.) It might be better to look at other explanations for the Canterbury report, some metaphorical, in order to interpret what happened there.

The moon we see from the earth exhibits constant features: The unaided eye's man or rabbit at full moon always looks pretty much the same (though "he" or "it" may look upside down to a moon watcher in the Southern Hemisphere). This fact may lead to a frequently held misconception: At first, you might think that it is because the moon is not rotating on its axis like the earth does.

But this is incorrect. In fact, the moon must be rotating in a special way for us to see what we see: synchronous rotation.

The moon rotates once in exactly the same time that it revolves once—one month. If it did not rotate, one side of the moon always would face a certain direction on the celestial sphere. Meanwhile, people on the earth, who face different directions on the celestial sphere at different times, would get to see one side of the moon sometimes and another side at other times. That does not happen, so the moon must be rotating, synchronously.

Synchronous rotation is due to the earth's gravity and the fact that the moon is not absolutely symmetrical. Of course the earth, in orbit about the sun, does not rotate synchronously.

We can accurately talk about the lunar nearside (the hemisphere facing us and a bit closer) and the lunar farside (the hemisphere facing away from us and a bit farther). However, some people speak of the "Darkside of the Moon." Now, that is a really good Pink Floyd album, but the phrase is silly as far as the moon is concerned. Each hemisphere of the moon experiences day and night equally. It just happens that we cannot see day or night occur on one of the sides (the farside).

One caveat: If we only could see the nearside of the moon, exactly 50 percent of the moon could have been mapped before robotic spacecraft first flew around the other side in the 1960s. In truth, our pre-space-age moon maps had greater coverage than that—about 59 percent.

During the course of half a day, an earthbound observer moves, due to the rotation of the earth, from a point one earth radius east of the moon-earth line to one earth radius west of the moon-earth line. The moon is close enough to the earth that the one-earth-diameter difference in vantage point through the night allows us to peer over the edge of the nearside. Similarly, due to the tilt of its orbit, the moon is sometimes north of the celestial equator; a northerly observer on the earth gets a glimpse of the farside over the lunar north pole. When the moon is south of the celestial equator, a southerly observer can do the same thing—over the lunar south pole. These are geometric, parallax effects.

More significant is the following: The moon seems to rock back and forth, as viewed from the earth. Sometimes we can see a little more to

Who Is the Man in the Moon?

The Man in the Moon is a product of the Christian era. To the Greeks and Romans, the moon was a female deity. He shows up in one of the most popular (and reprinted) astronomy textbooks of all time, *De Sphaera* (a fifteenth-century edition), by John of Holywood (thirteenth century; also known as Sacrobosco). The Moon Man likely still is older than that. European folk tales tell us that he was once an earthling. However, after committing some petty theft, he was caught and exiled to the moon. No wonder that the Man in the Moon often is pictured as looking rather unhappy with his status.

Selenologist Ewen Whitaker has catalogued other figures people have seen in the moon over the ages: an elderly man carrying a bundle of sticks, an elderly lady at a spinning wheel, two children carrying a bucket, and, of course, the rabbit. The Chinese rabbit sat pounding rice—an incongruous image to many of us. More recently than these imaginations, German schoolman Albert the Great (Albertus Magnus; circa 1200–80) saw a complicated scene: a dragon under a tree, with a man leaning against the tree.

From Scandinavia, we have the story of Jack and Jill, who, according to the rhyme, "went up the hill / To fetch a pail of water." If Jack (from the Swedish word for "increase"?) is seen in the first quarter moon, Jill (from the Swedish word for "decrease"?) is seen in the third quarter moon, and the full moon is Jack and Jill together, then the order of waxing and waning phases makes it true that "Jack fell down and broke his crown, / And Jill came tumbling after."

For over two millennia, one supposedly more naturalistic hypothesis concerning the features seen on the moon was that they were reflections of earthly features. In other words, the moon was a big mirror in the sky. The idea goes back to at least Clearchos of Soli, a Greek philosopher of around 320 BCE. It popped up again in the Middle Ages and appeared in a popular textbook, Robert Anglicus's 1271 commentary on *De Sphaera*. The argument against it, then as now, was that the moon rising in the east should reflect different terrestrial land masses than a moon setting in the west—a difference that is not seen. Yet, as late as 1570, an Arab cartographer drew a map of the earth, part of which looks a lot like a mirror-reversed image of the moon. *

*Philip Stooke, "The Mirror in the Moon," *Sky & Telescope* 91, no. 3 (1996): 96.

its west; other times we can see a little more to its east. While any effect that allows us to peek around the moon is called a libration, this rocking action is named physical libration.

One physical libration occurs because the moon travels at slightly different speeds at different places in its orbit (just as the earth does in its orbit about the sun). Its rotation, which takes place at a constant rate, must then be a little ahead at one point in the moon's orbit and a little behind when the moon is at the opposite point in its orbit. This situation results in a longitudinal libration, where by "longitude" I mean the longitudes (east and west) that we use to define locations on the moon.

There is also a physical libration called latitudinal libration. Again, here I refer to latitude (north-south) on the moon. The moon's rotation axis is tilted seven degrees with respect to the plane of its orbit about the earth. (It is a lunar equivalent of planetary obliquity.) Sometimes the moon's north pole is tilted toward us (and its equator is south of mid-disk); we can peer a bit over the pole to see some of the northern farside. Sometimes the moon's south pole is tilted toward us (and its equator is north of mid-disk); then, we can glimpse a bit over the pole to see some of the southern farside.

(Figures 8.1 and 8.7 both depict the full moon. Notice, though, that the photographs were taken at slightly different librations.)

THE NEW MOON IN THE OLD MOON'S ARMS

It is sometimes possible to see more of the lunar disk illuminated than its phase would dictate. Sometimes it is possible to see the rest of the disk, dimly. This means that in addition to seeing part of the lunar dayside (the phase, horn to horn), we also can see part of the lunar nightside. When this happens during the crescent phase, the phenomenon poetically is called "the New Moon in the Old Moon's arms."

If the moon produces no light of its own, from where does the aforementioned illumination come, if not from the sun? It comes from the only other bright object in the neighborhood: the earth. The earth reflects sunlight just as the moon does. The earth is pretty shiny (has a higher albedo than the moon), and it is big. It can be very bright. Sunlight striking the earth is reflected, and some of that reflected light strikes the moon (the whole disk facing the earth, not just the sunlit "phase"). This light is reflected again by the moon into our eyes. By now the light is pretty attenuated; much of it has been absorbed either by the earth or by the moon. Still, it is just bright enough to produce noticeable contrast with

FIGURE 10.2. Earthshine. The moon is a day and a half old. Courtesy of www.klipsi.com.

the pitch black of the night sky. The technical term for this light is earthshine, and it was first correctly explained by Leonardo da Vinci. Earthshine is about a thousand times fainter than normal "moonshine."

Earthshine comes from the dayside of the earth, obviously. It is too faint to see during the earth day; but just as part of the moon's disk facing the earth is lit by the sun and part is not (excluding new moon and full moon), so is the disk of the earth facing the moon normally partly sunlit and partly not. Indeed, if viewed from the moon, the earth will undergo phases just like the moon, and for the same reason. So it is possible to sit on the nightside of the earth watching the moon while it is illuminated by the dayside earthshine, originating perhaps some thousands of kilometers away. All you have to do is be on the side of the earth facing the very young or very old moon, at night.

Earth phases (as viewed from the moon) are one half-month out of synch with lunar phases (as viewed from the earth). For instance, a crescent earth takes place during a gibbous moon. So, not too many days after new moon (or just before new moon), not only is the waxing moon becoming glaringly bright enough to make seeing earthshine difficult, but less and less of the "waning earth" is sunlit and producing earthshine in the first place. It quickly becomes impossible to discern the earthshine.

If you call the "phase" of the earthshine the "old moon," then the "old moon" wanes while the proper moon phase waxes. Hence the name, "New Moon [or nearly so] in the Old Moon's arms" makes sense.

Earthshine varies ever so slightly due to meteorology on the earth's daylit hemisphere. More clouds mean a higher-than-average albedo— fewer clouds, less. So theoretically you should be able to say something about tomorrow's weather by looking at the earthshine the night before.

THE SOLAR PHOTOSPHERE

The only other regularly visible celestial object to exhibit features on its disk (like the moon) is one we already have discussed: the sun.

Normally the sun is a perfectly round yellow (nearly white) disk. Its color is a balanced mix of the colors visible to the human eye—red through violet—with yellow in the middle. Coincidence? Not really. Our eyes evolved under the light of this sun. If our biological ancestors had developed under the light of a star colored differently than the sun, its predominant color likely would be naturally selected as the midpoint of our visual range.

The sun may appear red at sunrise or sunset, but remember that this is caused by the earth's atmosphere preferentially scattering the sun's blue light and allowing only its red light to travel straight through from the source. (The same is true for the moon and stars, close to the horizon.) An incandescent gas, the sun is a bit darker, and redder, on its limb compared to its center. This is because the limb light is emanating mostly from the sun's outer, cooler layers (the photosphere). Even this inhomogeneity is almost impossible to make out with the naked eye.

When we see the sun in the absence of an eclipse (next chapter), we are looking into its photosphere. So what is there to see on the photosphere of the sun? Not much. And that is probably a good thing. With the light of a trillion, trillion one-hundred-watt light bulbs, the sun is fully capable of damaging the eyesight of one who gazes at it too long. Even a dog knows not to stare at the sun; there are no reports in veterinary medicine of a canine going blind from sun exposure. But there are cases in medical journals. Only "intelligent" people tend to override their own good, natural instinct to blink or look away when the sun becomes too much to bear. That said, it is seemingly safe to glance at the solar disk, for

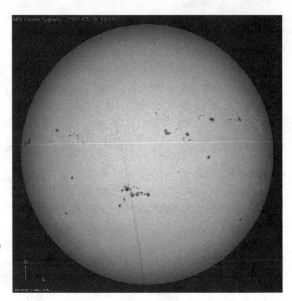

FIGURE 10.3. The solar photosphere (photographed through a protective filter). Notice both groups of, and individual, sunspots. Courtesy of Earth Observatory NASA.

a moment, through thin clouds, smog, fog, or dust; reflected in water; or at high-air-mass sunset or sunrise.

And what you might see, if you are lucky, is a photospheric feature called a sunspot. Sunspots appear jet black and seem to mar the gleaming photosphere. Most of these spots are too small for the naked eye to resolve, but there are exceptions. Also, sunspots often come in groups, which are easier to pick out than are individual spots.

The reason that sunspots look black is not because they do not produce any light. They are 1,500 °C or so cooler than their surroundings. They do not seem as bright as the incredibly bright photosphere. Our eyes cannot handle that much contrast, so they fool us into thinking that the spots are black. If we could somehow remove a sunspot from the sun, and see it by itself, it would look plenty bright.

Individual spots last from a few hours to a few months. They start out as small "pores" in the sun. Most of these die out. If such a feature survives, it grows into a full-fledged sunspot.

Science historian Judit Brody has cataloged pretelescopic observations of sunspots. All of these sightings must involve inordinately large sunspots. The earliest written description of a sunspot is on a Babylonian tablet made more than three thousand years ago. Aristotle's pupil Theophrastus (circa 371–287 BCE) wrote of a "black mark" on the sun. But we then must leap forward a thousand years to Einhard (circa 770–840), biographer of Charlemagne, who mentions a black spot seen on the sun. The year was 807.

A Gift from the Sun?

The sun produces a stream of subatomic particles that intersects the earth. It is named the solar wind. Described that way, the solar wind sounds benign. However, fast-moving "subatomic particles" equate to radiation, which can be harmful to life. Fortunately, the earth's magnetic field traps and deflects much of this material.

But not all. Occasionally, and unpredictably, a powerful burst of radiation, called a solar flare, escapes the sun. Like most temporary phenomena associated with the sun, flares are more common at sunspot maximum. The radiation they produce may overcome the barrier of our magnetosphere and hit the earth, especially near its Achilles' heels, the magnetic poles in the Arctic and Antarctic. Radio communication is affected. Electrical power is interrupted. Blackouts may occur.

I am happy to report that the earth has a second line of defense, preventing these side effects much of the time. The solar flare first strikes the earth's atmosphere, which absorbs most of its energy. Sometimes we see the air do battle on our behalf: The energy of the incoming radiation expends itself in a high altitude light show called the aurora. (The aurora also goes by the name of "northern lights" or "southern lights"; this is because it is most often seen in the north, by those living at high latitude in the Northern Hemisphere, and in the south, by those living at high latitude in the Southern Hemisphere.) Insofar as it happens within our own atmosphere, the aurora is not a celestial phenomenon. However, it is *caused* by one.

The aurora may appear as an ethereal curtain of moving, colored light. It is a lovely gift from the sun, but let us not forget its sinister side.

Western cosmology before the Renaissance stressed the unchanging nature of the heavens. Were sunspots unseen because they were not "supposed" to be there? On the other hand, the ancient Chinese were interested especially in celestial *change* (because of its significance to their astrology). They documented sunspots starting in at least 165 BCE.

In the nineteenth century, a German pharmacist and amateur astronomer named Heinrich Schwabe (1789–1875) took to counting the number of sunspots he could see on the sun at a given time. He did this for many years. He found that, on the average, there are lots more spots in some years than in other years. Furthermore, this rise and fall in the number of spots turns out to be periodic. There is a maximum number of sunspots every eleven years (approximately). Likewise, in-between, there is a minimum number of sunspots every eleven years (approximately). At sunspot maximum, there may be as many as one hundred spots visible through a telescope at one time. Some of these even may become discernible to the naked eye. At minimum, one may see no sunspots regardless of how one looks. During a particularly long sunspot minimum starting in 2007–9, few sunspots were seen for more than one entire year.

The eleven-year cycle is approximate and varies some, but it is regular enough to refer to the phenomenon as the sunspot cycle. It turns out that the occurrence of many solar phenomena is associated with the sunspot cycle, such as magnetic storms that interfere with power grids and radio communication on the earth.

Early in the sunspot cycle, look (briefly! See below.) for spots near the sun's equator (middle). Later in the cycle, look at higher latitudes (near the sun's north and south poles, at the solar limb).

It is the sun's magnetic field that causes sunspots. As the sun rotates (with a period of about thirty days), the field gets twisted up and forms a bottleneck to energy trying to escape. The details are still far from clear.

To reiterate, do *not* stare at the sun for any length of time. It will not help to wear sunglasses. They likely filter out certain colors of light, but not necessarily those which harm your eyes—these colors are invisible. The sunglasses simply may cause your pupils to open more widely and let in more harmful rays.

Never use a telescope or binoculars to look at the sun. These optical aids gather more light than your naked eye does. The increased amount of light that they funnel into your pupil may damage your eye severely. In the case of a large-aperture telescope, the damage may occur too quickly for you to react. As the nerves within your eyes do not respond to pain, even with a smaller instrument, you may not be able to feel your burning retina.

11

Eclipses

How does the Sun get its hair cut?
Answer: Eclipses it!

From the wrapper on a piece of Laffy Taffy candy

The word is overapplied, but I just cannot help but use it: Eclipses are *awesome*.

SOLAR ECLIPSE

My first solar eclipse was not a total solar eclipse where I lived. (I will explain the difference later.) Nevertheless, I remember it well. It was March 1970. I was a kid, lying on my back in my yard. I think the sight greatly affected my decision to become an astronomer.

Some may remember a well-publicized solar eclipse in the United States again, in the year 1979. But I did not see a total solar eclipse until 1991. I knew there was going to be one in the United States: That state was Hawaii, so what was the downside? If the weather was poor, I only would be stuck in Hawaii. Not bad.

I did see that eclipse. This was ironic because Hawaii happens to be the site of some of the world's great observatories—which were largely clouded out, elsewhere on the Big Island.

I remember that eclipse, and each one I have seen since, as vividly as my first. I also remember the curious human behaviors that eclipses bring forth: For example, during the 1994 South American eclipse, I recall the gunshots that rang out at mid-event. Was somebody trying to scare away the eclipse monster?

In 1998, I took my students to Antigua for a total solar eclipse. (The Caribbean in our northern winter was not a hard sell.) In 1999, I saw the show on the Romanian seashore with hundreds of strangers. In the 2000s, I saw it again from Turkey and from Mongolia with a few friends.

FIGURE 11.1.   A total solar eclipse in western Egypt, March 29, 2006. The sun's photosphere is completely obscured. (This is a composite photograph.) Photo by Gregory E. Morgan.

My record is modest. I know a gentleman who has traveled to see twenty-seven total solar eclipses.

Why all this time and expense? Eclipses are, I am convinced, the most spectacular occurrence in our sky. I think that few people have ever witnessed a total solar eclipse and been unmoved. There are stories of clueless motorists merely turning their lights on and proceeding down the highway, but I find them hard to believe.

What is it that tantalizes us eclipse junkies? First, the dark silhouette of the moon encroaches on the brilliant sun. When half of the sun has been extinguished from our day, we notice dullness to the blue sky. Minutes later, the quality of light gives everything below a gray, metallic cast. The western sky is clearly darker now, regardless of where the sun is in the sky. This darkness, the shadow of the moon, expands and rushes toward us, always eastward. Once the sun has been reduced to a mere slit, we may be lucky enough to see cells of air above us as shadow bands racing across the ground. It is as if we stand on the bottom of a fish tank, and a giant has stirred the waters. A total solar eclipse has begun.

A tour d'horizon reveals 360 degrees of sunset colors. The bright disk of the sun disappears in a flurry of one or more sparkling Baily's Beads, glints of sunlight seen through lunar valleys and between lunar moun-

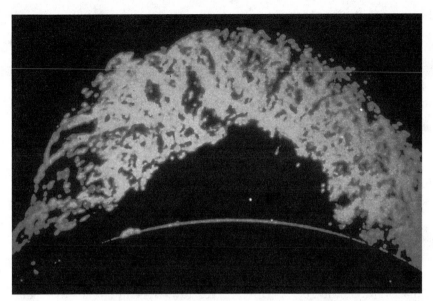

FIGURE 11.2. A solar prominence. Courtesy of Collection of Commander James O'Clock (NOAA Corps).

tains on the rough limb profile of the moon. During the few minutes of total solar eclipse, all is darkened. The temperature goes down. The planets and, rarely, bright stars can be seen, even though it may be the middle of the day. Birds return home to roost. The only illumination is from the thin, red chromosphere and the amorphous corona, the faint outer layers of the sun, normally drowned out in the glare and blue sky produced by the now absent photosphere. (See figure 11.1.) Long creamy streamers radiate from the corona, several sun diameters long. The sun is no longer a circle. The abrupt end of totality is signaled again by Baily's Beads or, perhaps, another single, bright diamond-ring effect (also potentially visible as totality began).

Occasionally, red tongues of what looks like flame are seen erupting from the solar chromosphere during the total phase of the eclipse. They are called prominences. It is no wonder that, for ages, people thought that the sun was on fire.

A total solar eclipse is a visceral experience.

Only nine total eclipses of the sun were visible in the continental United States during all of the twentieth century (though, coincidentally, in the thirty-eight years between 1594 and 1632, eight total solar eclipses were visible from what would one day be the United States or Canada, a statistical oddity). During the Middle Ages, London went 837 years without a total solar eclipse. On the other hand, southern New Guinea witnessed a total on June 11, 1983, and also on November 23, 1984—a wait of only one and a half years. The next total solar eclipse visible from anywhere in the United States will be in the year 2017. It is estimated that only one in ten thousand people worldwide have experienced this exotic natural phenomenon.

To say what an eclipse is, I must first say what it is not. Do not confuse phases of the moon with eclipses. Eclipses are rare compared to phases.

A complete cycle of lunar phases occurs every synodic month. Remember that new and full moon occur when the three bodies, earth, moon, and sun, form a syzygy. It does not have to be a perfect lineup. In fact, most of the time, the moon is a little above or below the line running from the center of the sun through the center of the earth.

What happens when these bodies are perfectly lined up? For instance, sun, moon, earth? Imagine shooting an arrow and striking all three through the middle. Then, and only then, does the moon physically block light emitted by the sun from reaching the earth. It is a total solar eclipse.

The moon occludes the sun. So? Objects pass in front of other things in the sky all the time. The moon covers up a star, but nobody notices one star's worth of light missing. A planet, such as Venus, crosses the solar disk, but it is a barely detectable dark spot. These are unspectacular events because the bodies involved are of such disparate apparent sizes.

However, by a wonderful coincidence, the sun and the moon appear to be the same size in the sky. The moon is about the extent of a pea held at arm's length: half a degree. We tend to think that the sun is bigger (because it is so much brighter?)—yet, in reality, you can blot it out with that same pea.

This coincidence happens nowhere else in the solar system. The moon is 3,476 kilometers in diameter. If it were only 260 kilometers smaller, we would never see a total solar eclipse. It is nice that such a thing as an eclipse happens on the one planet where there are people around to see it.

Our opportunity will not last forever. The moon is working its way away from the earth: Its orbital radius increases by 3.8 centimeters every year. Eventually, the moon never again will appear large enough to cover the sun. The final total solar eclipse will take place—hold your breath—1.2 billion years from now.[1]

Back to the present: Why does an eclipse not occur every month? Recall that the moon's orbit is inclined to the plane of the earth's orbit around the sun—the ecliptic. ("Ecliptic" . . . "eclipse": Get it?) There are only two places in the moon's orbit where it is in the plane of the eclip-

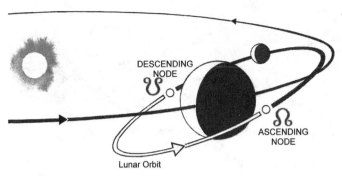

FIGURE I I.3.   The nodes of the lunar orbit. Not to scale.

tic (the two places where it crosses the ecliptic). These points are called nodes. "Node" comes from the Latin word for "knot," as in tied together for weaving. Nodes are not things, they are places—most of the time they are empty places.

For a total solar eclipse to occur: It must be new moon, and at the same time the moon must happen to occupy one of the nodes of its orbit. As the time it takes the moon to go from node to node to node is different and unrelated to the time it takes to go from phase to the same phase again, for these two events to occur simultaneously requires yet another coincidence. If the conditions are not met, there will not be an eclipse. Or if the moon is just a little off from a node point, there will be a partial solar eclipse (like my first eclipse experience)—and the disk of the moon will "bite out" part of the solar disk, though not all of it.

A partial eclipse also proceeds and follows a total solar eclipse. Such an eclipse begins with first contact. During the subsequent partial phases, the moon's shadow makes the sun look like Pac-Man. You know those little circles that appear underneath a leafy tree when the sun is shining? Those actually are pinhole-projected images of the sun. During a partial solar eclipse, they appear as crescents.

Recall that shadows seem to have two parts: a darker, inner part—the umbra—and a lighter, outer part—the penumbra. After first contact, we (the eclipse observers) are in the moon's penumbra. Then, if the eclipse is total, at second contact we are now in the moon's umbra. Again, if total, at third contact we return to the penumbra. At fourth contact we are out of the moon's shadow altogether.

People within the path of totality, the width of the moon's narrow umbra, only experience a second and third contact. Adjacent to the path, observers remain in the penumbra and only experience a partial eclipse throughout. Maximum totality duration is to be found in the middle of the eclipse path, on the centerline.

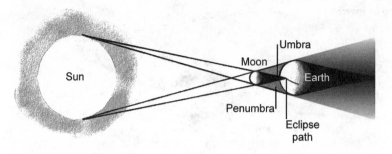

FIGURE 11.4. The circumstances for a total solar eclipse. Observers beneath the umbra (on and near the centerline) see a total eclipse. At the same time, observers beneath the penumbra see a partial eclipse. Not to scale.

This distinction was quite graphic during the 1925 solar eclipse viewable from New York City: Witnesses say that the "edge" between the moon's umbra and penumbra fell somewhere between Ninety-Fifth and Ninety-Seventh Street in Manhattan. (Some scientists intentionally station themselves—not on the centerline of the eclipse path—but on the predicted border between totality and partial eclipse; they hope that whether they see a total eclipse or not will provide data on any possible change in the size of the sun over time.)

Incidentally, the moon is not always just the right apparent size to cover the sun. Remember that the moon, like everything else, travels in an elliptical orbit. There is a point in this orbit where it is closest to the earth, and a point where it is farthest. When farthest, at apogee, the moon looks smaller than it does when it is closest, at perigee. Meanwhile, the apparent size of the sun changes slightly between the earth's closest approach to it (perihelion) and the earth's farthest recession from it (aphelion). These two effects can conspire to, during certain eclipses, make the sun appear a little larger than average and, at the same time, the moon appear a little smaller than average. In such an eclipse, the moon does not appear big enough to cover the solar disk, even if it moves directly over the middle of it. The result is not totality but an annular eclipse. (Total and annular eclipses considered together are called central eclipses.)

Think of an annular eclipse as a negative moon shadow. Instead of the tip of the shadow cone (theoretically) reaching into the earth, it ends short of the earth's surface. In the midst of an annular eclipse, the sun looks like a big ring ("annulus") in the sky. There are even central eclipses that are hybrid: They are annular at the start of their paths, total in the middle, and then annular the rest of the way.

Eclipse magnitude refers to the percentage of the sun that is eclipsed: One hundred percent or more is a total eclipse, less than 100 percent is

FIGURE 11.5.   An annular eclipse at sunset, January 4, 1992. Photo by Hiram Clawson.

a partial eclipse. Few people will notice an eclipse of less than eclipse magnitude 0.7, unless it is pointed out to them.[2]

A total eclipse (eclipse magnitude 100 percent or more) is the only time it is safe to look at the otherwise wince-producing sun. (The corona is just about as bright as the full moon.) Remember that the sun normally is so intense that it can damage your eyesight if you stare at it. Sometimes the damage is permanent. It may be counterintuitive to think that a total solar eclipse is the only safe time to view the sun directly—the bright photosphere is blocked out between second and third contact—because of the myth that eclipses are dangerous. It is the sun that is dangerous, not the eclipse.

There are sad stories—sad to me, anyway—of school children locked up during an eclipse to save them from the (imagined) rare and mysterious "eclipse rays." Such a situation might arise because we would not ordinarily think of peering at the sun. However, just before a total eclipse, something particularly interesting happens to the sun—a partial eclipse. Unfortunately, a partial eclipse is the same as no eclipse, as far as eye safety is concerned. Each tiny fraction of the sun's disk is as bright as any other, and even one sliver is intense enough to cause harm. Yet, because of the interest in the changing appearance of the sun as the eclipse progresses, there is the inclination (ironically, among so-called higher species like humans) to override common sense and stare at the sun far too long at these times. Herein lies the danger. Nevertheless, there is noth-

FIGURE 11.6. These Mongolian boys use proper black-polymer-resin filters to watch August 1, 2008's total solar eclipse before second contact and after third contact. They safely removed their filters in between. Photo by Yuliana Ivakh.

ing different about the quality of the sunlight before, during, or after an eclipse.

To reiterate my warning: *do not* observe the uneclipsed sun *directly* without proper precautions, such as filters approved by your local astronomer. (Some commercial filters work great, others do not.) Viewing the sun's image, projected through a pinhole, provides a way to observe a solar eclipse *indirectly* and obviates the need for any eye projection.

LUNAR ECLIPSE

So far I have emphasized one kind of eclipse, a solar eclipse. There is another kind of eclipse. It is a lunar eclipse. During a lunar eclipse, it is the moon that becomes dark, not the sun. As you might guess, lunar eclipses are a nighttime event. And because the moon is so much fainter than the sun, a lunar eclipse might be missed: I recall walking around Saigon, Vietnam, during a total lunar eclipse; there were hundreds of people wandering the street, going about their evening's business, sadly oblivious to the light show above them.

FIGURE 11.7. A lunar eclipse, June 4, 1993. Photo by Andy Steere.

Let us look at what happens when the moon is on the other side
of the earth. This time, the moon is not in the way, the earth is. If (1) the
syzygy is perfectly lined up: sun, earth, moon and (2) the moon is at one
of its nodes (a necessary condition for all eclipses)—then the shadow
of the earth falls on the moon. Sunlight does not reach the moon. Because the moon only shines by reflected light, it looks dark. This is all the
more noticeable because the phase of the moon at this time is normally
full. The event is called a total or partial lunar eclipse, depending upon
whether the moon is entirely in the earth's shadow or not.

The lengths of lunar eclipses are determined by the time it takes the
moon to pass through the earth's shadow. They last hours, not minutes.

The moon never completely disappears in a total lunar eclipse. Sunlight is refracted around the earth's atmosphere—red light the most. It
is this weak light that gives the eclipsed moon an eerie, ruddy glow. (Figure 11.7 shows a fine example.) The exact appearance will depend on
the state of the atmosphere: whether there are clouds on the terminator,
how much dust happens to be in the air, and that sort of thing.

Remember Aristarchus from chapter 8? He estimated the diameter
of the moon, compared to the diameter of the earth, by eyeballing the

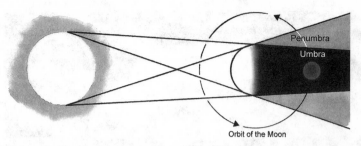

FIGURE 11.8. The circumstances for a total lunar eclipse. The moon is passing through the earth's umbra. If it passed only through the penumbra, the eclipse would be difficult to discern. Not to scale.

size of the earth's shadow during a total lunar eclipse. There was some lucky guessing involved, but he came up with a number not too far off the modern value.

Here is a riddle: Lunar eclipses are slightly more rare than solar eclipses. Nonetheless, you are more likely to see a lunar eclipse. Why? In a solar eclipse, the shadow cone is very narrow. The total eclipse is only visible in the spot where the cone touches the earth. Elsewhere, only a partial eclipse is seen. You have to be in this spot, sometimes only a few kilometers or tens of kilometers across, to see the total eclipse. (Actually, the spot is turned into a band—the path—across the earth, by the earth's rotation during the eclipse.)

During a lunar eclipse, everybody for whom it is nighttime can see the eclipse. (Full moon rises at sunset and is above the horizon until sunrise, remember?) That is half of the world.

So if you are staying in one place, it is unlikely that it will be the right place to see any given total solar eclipse. Your odds are about one in four hundred years. Total solar eclipses are for travelers. But in the case of a lunar eclipse, it is fifty-fifty: If it is night, you see it. There is a window of opportunity of a little more than twelve hours. About 29 percent of lunar eclipses are total, about 35 percent are partial, and about 36 percent are penumbral lunar eclipses. (The latter are admittedly so subtle that they are hard to detect with the naked eye.)

December 2010 saw the total eclipse of a blue moon. What are the odds? I do not know. However, given that the event occurred, you and I had a 50 percent chance of seeing it.

PREDICTING ECLIPSES

The moon steals the sun. That was a big deal to peoples in the past. (It still is to us.) A reason for past concern/superstition about eclipses was their seeming unpredictability. What governs the circumstances of an eclipse?

Remember the nodes? They are the points at which the moon's orbit intersects the ecliptic (the earth's orbital plane). When the moon appears to travel north, "up" through the ecliptic, it is at ascending node. When it appears to travel south, "down" through the ecliptic, it is at descending node. The line of nodes is an imaginary line connecting these two points. This line slowly shifts around the earth. So the time from the moon's presence at ascending node to again at ascending node, or descending node to descending node, is not exactly one synodic (or sidereal) month: It is 27.21 days, the nodical month (draconic month). Eclipses must happen when the moon is at a node (so that the earth, moon, and sun are all in the same plane). It must, at the same time, be either full moon or new moon, which are specific times of the synodic month. (It depends, of course, on whether we are talking about a lunar or solar eclipse.) This is a rare coincidence—but it would be fairly predictable if the nodes were always in the same place on the celestial sphere. Then, if the full or new moon approached this special celestial spot, an eclipse could be prognosticated.

Because the nodes appear to drift through the ecliptic, it is difficult to keep track of their locations and to predict eclipses. Or where the path of the eclipse will be (in the case of a solar eclipse, where an imaginary line going through the center of the sun and moon will intersect the earth).

Moreover, in the eclipse-of-the-sun scenario, it is harder to predict the type of solar eclipse, because the apparent size of the moon is always changing. A line connecting the moon's apogee point and perigee point is called the line of apsides. This line slowly rotates around the earth; the time between apogee and apogee, or perigee and perigee, is slightly different from the sidereal, synodic, or nodical month. It is 27.55 days—the anomalistic month. The moon appears bigger when near perigee, smaller when near apogee; this difference in apparent size could mean the difference between an annular eclipse and a total solar eclipse. The apparent size of the moon affects the width of the total solar eclipse path, too. (If the moon is closer to us, it casts a bigger shadow onto the earth.) The time of the anomalistic month may determine whether you are on the eclipse path at all.

If the moon were at its average distance (or farther) from the earth all the time, there would never be a total solar eclipse. From this fact you may deduce correctly that annular eclipses are a bit more frequent than are total eclipses.

These different intervals of time also affect the duration of eclipses.

Thus, in order to predict an eclipse (and what that eclipse must be like), one has to keep track of five variables: the times of the sidereal, synodic, anomalistic, and nodical month, as well as the time of year—not an easy job without a calculator. The idea that Stonehenge, or any other prehistoric monument, was designed to predict eclipses seems farfetched.

Still, there are some tricks that make eclipse prediction easier. Solar eclipse: The sun must pass through a lunar node twice a year, so there must be two eclipse seasons during which an eclipse might take place. (Eclipse seasons are independent of the real seasons.) You can have at least a partial eclipse of the sun any time when it is within fifteen and one-third degrees of a node. One can be alert for an eclipse during the days between the eclipse limits, and not worry about one the rest of the year.

The earth's speed varies during its elliptical orbit about the sun, but the time between eclipse limits is going to be between thirty and thirty-seven days (depending upon the dates). In fact, there must be at least one solar eclipse each eclipse season.

The synodic month is less that the length of the eclipse season: Notice that it is possible to have two solar eclipses in a single season if one occurs early and the other late. That yields four as the possible number of solar eclipses per year—or five if the first eclipse happens before January 18 (January 19 in a leap year). This is because our eclipse year, from eclipse season to eclipse season to eclipse season, is shorter than the tropical year. The nodes shift westward a little each year, so the eclipse year (defined as twelve nodical months) is only 346.62 days. A similar calculation can be made for lunar eclipses.

Seven is the maximum number of eclipses (any kind) that may occur in one Gregorian year. The sum is always three eclipses of the moon and four of the sun—or two of the moon and five of the sun. There were five solar eclipses in 1935; what is more, there were two lunar eclipses that year. This kind of record-setting year will not happen again until 2485.[3]

We can group together the related, overlapping, and (sometimes) conflicting societies that dwelled at the confluence of the Tigris and Euphrates Rivers during the first several millennia BCE, as the Mesopotamians. Here was the so-called Cradle of Civilization, home to the first great city builders. And it was here, fittingly, that people noticed and recorded that there is a pattern to eclipses.

The Mesopotamians found a cycle of eclipse activity that repeated every 223 synodic months. This is the period of time at which the synodic month, nodical month, and anomalistic month are nearly commensurate with each other. It is called the saros cycle. There are approximately seventy-one solar eclipses each saros (an interval of 6,585.3 days). Similar eclipse geometry happens at the same time each saros. For instance, eighteen years, eleven days after one lunar eclipse, there is almost always another. If, for instance, a total solar eclipse occurs, the next one, a saros later, also is most likely to be total (or maybe annular)—and in the same part of the celestial sphere. Only that fraction of one-third of a day (one-third earth rotation) guarantees that each eclipse will occur one-third of the way around the earth from the previous one. This usually does not push a long total lunar eclipse out of view, but it does the brief total solar eclipse. You have to wait three saros cycles to experience the same eclipse at the same longitude.

And even then: That "same" eclipse will occur a thousand or so kilometers north or south of the last "same" eclipse. Each successive eclipse in a given saros will be deeper (meaning less partial and more likely total) during the first portion of the saros. The opposite will happen during the second portion. (The saros is nineteen eclipse years.) After about 1,300 years, an individual saros cycle (a family of similar eclipses) "dies."

What the Mesopotamians actually did circa the eighth century BCE was this: Apparently they figured out that eclipses tend to be separated by five or six full moons—though an eclipse was not guaranteed at that interval.

The saros is a good model of eclipse behavior, but not exact. Predictions based upon it alone will eventually fail. Still, it is a tribute to ancient record keeping that the saros was discovered as early as it was.

Ultimately, the answer to the question "who first predicted an eclipse?" depends upon our criterion for success. Societies the world over have made the attempt. More than one have succeeded in establishing eclipse probabilities: when an eclipse was more or less likely to occur.

I have written about predicting the event itself. What about eclipse duration? During a total solar eclipse, the shadow of the moon travels along the eclipse path on the earth at 3,380 kilometers per hour (average). Yet at the equator, that speed is only 1,670 kilometers per hour. This is true because the surface of the earth is traveling faster at the equator (in the opposite direction of the shadow) and canceling out some of the shadow's speed relative to the observer. This surface speed is simply due to the diurnal rotation of the earth about its axis: A point on the equator must travel a forty-thousand-kilometer circle in the same time that a point near one of the poles travels a much smaller circumference.

So eclipses near the equator are longer than eclipses in the far north or south. To make the eclipse a bit lengthier still, the moon/sun should be directly overhead. This, too, is possible in the Tropics. Nonetheless, the longest total solar eclipse anywhere will last no more that seven and a half minutes.

ECLIPSES AND PEOPLE

So out of the ordinary are eclipses that they have long been considered omens, for good or ill. Venerated or feared, they rarely have been ignored. Entire books have been written on humankind's reactions to eclipses. I will provide here just a taste.

It is impossible to say what people first thought of eclipses. That report is lost to prehistory. However, we do know that *Homo sapiens* have a long history with eclipses. The circumstances of 13,200 lunar and solar eclipses all over the globe have been calculated, just between the years 1207 BCE and 2161. We note that the Chinese were writing about something that sounds like the corona of the sun, only seen during eclipses, as early as 1300 BCE.

Too, there is the famous story from around this time of two court astronomers, Hi and Ho, who were supposedly executed for failing to predict an eclipse—a much harder feat than observing one. This tale is, though, almost certainly apocryphal.

Still, as late as 840, an emperor is said to have died of fright during an eclipse. This is understandable in view of the Chinese myth that an eclipse occurs when a dog devours the moon. As recently as 1948, an election was postponed in South Korea on account of an eclipse.

The Bible may record an ancient eclipse: "And on that day, says the Lord God, I will make the Sun go down at noon, and darken the Earth in broad daylight" (Amos 8:9). Historians speculate that this passage is in reference to the eclipse of June 15, 763 BCE, which was total in Old Testament lands.

Elsewhere in the Bible, what sounds like a solar eclipse accompanies the crucifixion of Christ; however, no literal eclipse circumstance matches fully with other associated events in the New Testament story. On the other hand, Acts 2:20 ("The Sun shall be turned into darkness, and the Moon into blood, before that great and notable day of the Lord comes") does sound a lot like a lunar eclipse. There was such an eclipse on April 3, 33 AD, but it was not visible from Jerusalem.

Farther west, we have this poem by Archilochus:

> Nothing there is beyond hope,
> Nothing that can be sworn impossible,
> Nothing wonderful,

Since Zeus,
Father of the Olympians,
Made night from midday,
Hiding the light of the shining sun,
And sore fear came upon men

Archilochus lived during the seventh century BCE. A total eclipse of the sun was visible from Greece on April 6, 648 BCE.

The historian Herodotus claims that, during a war between the Medes and Lydians, the great scientist Thales predicted an eclipse. (This seems unlikely; perhaps he warned of the possibility of an eclipse in the manner of the Mesopotamians?) Anyway, both sides wanted peace, and this was as good an excuse as any to call off the war. The story can be associated with the eclipse of May 28, 585 BCE.

Helicon of Cyzicus's supposed prediction of a May 12, 361 BCE solar eclipse, like Thales's, also can be attributed to a misunderstanding. It is easy to believe that stories of a famous scholar *observing* an eclipse evolved into stories of his *predicting* an eclipse.

Filled with more verisimilitude, I think, insofar as it does not involve a prediction, is the story of the lunar eclipse of August 27, 413 BCE. It, too, took place during a war—this time between Athens and Syracuse. The Athenian commander viewed the eclipse as a bad omen. He retired from the battlefield for a full month, thereby giving Syracuse time to regroup. Syracuse won.

An exciting modern discovery in archaeology is a set of metal gears and dials remarkably preserved in a two-millennia-old Greek shipwreck. It is named the Antikythera mechanism (after a nearby Mediterranean island). As far as we now know, there was nothing else like it built for a thousand years. The Antikythera mechanism seems to be a sophisticated, mechanical calendar. New analysis techniques reveal what looks like a saros eclipse–prediction dial.[4] This one-of-a-kind artifact still is being studied.

While the Antikythera mechanism sat on the bottom of the sea, time floated by. The cause for eclipses became known. Moreover, Aristotle (384–322 BCE) pointed out that, because the earth always produces a round shadow on the moon, the earth must be a sphere.

However, that did not mean that everybody knew what was going on. Roman citizens still might be heard making a great clamor during a lunar eclipse, the purpose of which was to frighten away the great wolf that was eating the moon.[5]

Roger of Wendover (died 1236) chronicles the early-day eclipse of

FIGURE 11.9.   The Antikythera mechanism. It now resides at the National Archaeological Museum, Athens. Photo by Rien van de Weygaert, Groningen, the Netherlands (www.astro.rug.nl/~weygaert/antikytheramechanism.html).

May 14, 1230: He tells of the sky becoming so dark that laborers, who had commenced their morning's work, returned to their beds to sleep—only to restart the day hours later, after the eclipse had ended.

The nature of the corona was still a subject of speculation. It had been documented in print since the tenth century, but what exactly was it? (The word itself was not coined until 1803.) In the seventeenth century, Johannes Kepler (1571–1630; arguably the first modern astronomer) correctly concluded that the corona was actually part of the sun, and not, say, of the moon, nor was an optical illusion produced by an eclipse.

Eventually, it became possible to predict eclipses accurately. (In the Western Hemisphere, the Maya probably could crudely foretell eclipses by the third or fourth century.) In 1504, Christopher Columbus was marooned in Jamaica, awaiting resupply. He told the natives that if they did not provide him and his crew with food in the meantime, he would take away the moon. Columbus was able to demonstrate the power of his threat since he knew that a lunar eclipse was to occur on February 29, and he timed his pronouncement accordingly. The ruse worked. This was one of Columbus's more "pleasant" interactions with New World natives.

History Loves a Good Eclipse

Modern historians appreciate records of eclipses. These help to pin down dates. For example, Haco IV, king of Norway, is supposed to have attempted an invasion of Scotland in 1262. In the annals of that military expedition, we read that "a great darkness drew over the Sun, so that only a little ring was bright around his orb." This is about as clear a description of an annular eclipse as we get. Now the astronomers weigh in: They compute that an annular eclipse did take place, visible from that part of the world, but in the year 1263. The history books are corrected.

By the way, Haco was defeated. That is why the Scots do not today speak Norwegian.

Skipping forward to more recent times, eclipses were standard observing fare for observational astronomers, who wished to keep track of the exact location of the moon. The coordinates of the moon were of practical value to mariners, who used its position on the celestial sphere to determine longitude before the advent of reliable sea clocks. Timing a total eclipse allowed placing the moon precisely at the exact location of the sun.

A lunar eclipse was of even more direct use: Because such an eclipse is a simultaneous event over large areas of the earth, comparing the time of its occurrence to that printed in an almanac, written for a known location, allowed calculation of the time difference, and hence longitudinal difference, between your location and the almanac's.

12

Placing Planets

A good friend of mine
Studies the stars
Venus and Mars
Are alright tonight

PAUL MCCARTNEY, "Venus & Mars," 1975

While our eye and brain insist on equating brightness with size, the planets are no more disklike than stars are to the naked eye. The fact that these five special objects (the planets visible without a telescope) were noticed long ago and thought to be exceptional was based upon their motion on the celestial sphere. They were not "fixed" stars, well behaved because they stayed in constellations. They appeared to "wander" through the zodiac. Indeed, "planet" meant "wanderer" in Greek.

PLANET PATHS

The apparent motion of the planets in our sky is more complicated than that of the stars, sun, or moon. In the case of the stars, they are so far away that the orbital motion of the earth about the sun inconsequentially affects our point of view. Only the diurnal rotation of the earth (plus precession) causes them to appear to move at all. In the case of the sun's apparent motion, we did have to take into account both the rotation and revolution of the earth. In the case of the moon, it was the earth's rotation and the orbital motion of the moon itself about the earth. Still, the rate of the sun's apparent motion (due to the earth's revolution about the sun) and the moon's motion (which in reality is due to the moon's revolution about the earth) are fairly constant, helping to keep things simple.

Planets, though, are close enough that in charting their apparent motions in the sky we must take into account the earth's rotation, the earth's revolution, and—independently—the unique and sometimes more variable revolution of the planet itself about the sun. Moreover, each planet

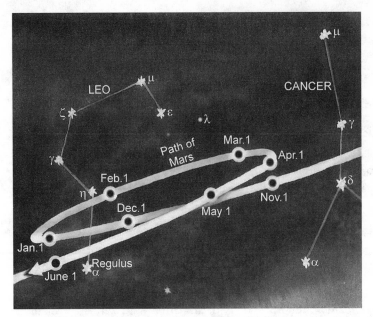

FIGURE 12.1. The path of a planet can appear confusing. Most of the time—but not all—
this group of planets travels eastward (right to left in the figure), in the prograde direction.

has a different orbital path around the sun: These orbits are noticeably elliptical, with each having a different size, eccentricity, inclination,[1] and orientation. Things get a bit messy, and, for this reason, you will not see me discussing the position of the planets (their rising and setting points on the horizon, their altitudes as they culminate, etc.) in great detail.

An aside: Just because we are skipping the subject of risings and settings does not mean that everybody has done so. In (about) ninth-century Mexico, the Maya aligned a building at Uxmal, Yucatán (now called the Governor's Palace), to the most southerly rising point of Venus. We believe that this association with the planet is intended because the inscriptions on the edifice are a veritable billboard advertising "Here is Venus!" We will revisit the Maya later in the chapter.

Still, there are some general things that can be said about the motions of the planets. First, most planets spend the great majority of time moving predominantly west to east on the celestial sphere. (Unlike with the sun or moon, there are a couple of exceptions to this statement in regard to the planets—look at figure 12.1—I will mention these later.) We call this direction of planet movement prograde. This is because all planets, including the earth, revolve about the sun in the same direction: counter-clockwise as viewed from the direction we call north. Second, because the orbital planes of all the planets are similar, Mercury, Venus,

Mars, Jupiter, and Saturn all can be found near the ecliptic. (Our solar system is pretty flat.)

Each planet has (different) sidereal and synodic periods, just like the moon. Mercury travels fastest around the sun among all the planets; Venus is second swiftest. Both have sidereal periods shorter than that of the earth, while the other planets all have longer sidereal periods than the earth. Because Venus is the closest planet to the earth, and is traveling at a speed around the sun not much greater than ours, it takes Venus a long time to lap the earth. The synodic period of Venus (and Mercury, too) is much longer than its sidereal period.

In the case of Mars, the next closest planet to the earth and one traveling only a little more slowly than the earth, it is the same: It takes the earth much longer than a year to overtake Mars.

The planets of the outer solar system, beyond Mars, move comparatively slowly against the celestial sphere. The earth catches up with them easily and their synodic periods are not much different from their sidereal periods.

INFERIOR PLANETS AREN'T BAD

We all have learned that the earth is the "third rock from the sun"—that is, it is third in order among the nested orbits of the planets. This simple fact results in there being two kinds of planets in our sky: It may seem rather obvious, but there are planets that orbit farther from the sun than we do here on the earth and other planets that orbit closer to the sun than we do. Yet this simple distinction largely governs how we see the planets in our sky.

The planets inside the earth's orbit are called inferior planets. There is nothing wrong with these planets. They are "inferior" only in the sense that their orbits are smaller than that of the earth's. Because of this, these planets are never far away from the sun in our sky. "Chained" to the sun, from our point of view, inferior planets can never be on the opposite side of the celestial sphere from the sun.

This geometry makes it harder to catch an inferior planet. If it is always near the sun, much of the time that the planet is above the horizon, the sun is also up. We cannot easily see the planet against the blue sky. If it is west of the sun, our best bet is to spy the planet after it has risen, but before the sun rises. Alternately, if the planet is east of the sun, we might spot it just after the sun sets, but right before the planet does. In both cases we are probably looking for the planet in twilight.

Both Mercury and Venus are "morning" and "evening" stars (names coined before the modern distinction between planets and stars). The names indicate the times of day we are likely to see them. Venus, planet "number two," gets angularly farther from the sun than does Mercury, planet "number one." Plus, it moves more slowly on the celestial sphere than does fleet-of-foot Mercury. These data, and the fact that Venus is a brighter planet than Mercury, mean that when somebody points out *the* morning or evening star, they almost certainly are indicating Venus.

When the angular separation between an inferior planet and the sun as seen in our sky is greatest, astronomers say that it is at maximum elongation. The maximum elongation of Venus (forty-eight degrees) is, of course, bigger than the maximum elongation of Mercury (twenty-eight degrees).[2] Every synodic period, an inferior planet reaches a maximum eastern and maximum western elongation.

Here is where our statement that planets appear to move mostly in a prograde direction is not so clear: As an inferior planet swings between elongations in our sky, it would seem that about half the time it should appear to move prograde (west to east) and about half the time it should appear to be moving retrograde (east to west). When the planet shifts from maximum western elongation through the far part of its orbit to maximum eastern elongation, though, it is traveling through a longer arc of its orbit than it does when it is traveling from maximum eastern elongation through the near part of its orbit, to maximum western elongation. So its prograde motion on our celestial sphere really does last longer than its retrograde motion.

I have described maximum elongation as a good time to see an inferior planet in the sky; which one depends on whether you want to be awake before the sun rises or after it sets. There also are two times when it is almost impossible to observe an inferior planet: when the earth, sun, and planet are nearly lined up. If an inferior planet and the earth are on opposite sides of the sun, as must happen once during each of the inferior planet's synodic revolutions, to look at the planet we must also look in the direction of the sun. You cannot make the planet out. The planet and sun are said to be in conjunction (lined up). Specifically, the inferior planet is at superior conjunction.

If an inferior planet and the earth are on the same side of the sun, as must happen once during each of the inferior planet's synodic revolutions, to look at the planet we must once again look in the direction of the sun. This lining up is called inferior conjunction.

Did I not use the word "inferior," in two different contexts, within just the last couple of pages? I am afraid so. We distinguish whether an inferior planet is at superior conjunction or inferior conjunction based

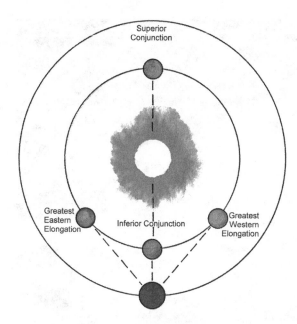

FIGURE 12.2. An inferior planet in four different geometries with respect to the earth and sun. Not to scale.

on whether it is farthest from the earth at the time (superior conjunction) or closest to the earth at the time (inferior conjunction). Either way, inferior conjunction or superior conjunction, this is a poor time to try to see the planet.

It takes Venus 220 days to go from superior conjunction to maximum eastern elongation. It then takes another seventy-two days to move from maximum eastern elongation to inferior conjunction. Similarly, it takes seventy-two days from inferior conjunction to maximum western elongation. It is another 220 days from maximum western elongation back to superior conjunction.

Here it gets interesting: Venus is invisible near superior conjunction from ninety to fifty-five days. Yet it is invisible near inferior conjunction from only nineteen days to a mere one day—or less. This means that it is theoretically possible for Venus to appear as a morning star and evening star on the same day.

Technically, you can see—barely—the planet Venus if it actually appears to cross the disk of the sun, too. (Mercury is too small and too far away from the earth for it to be visible, in transit, to the naked eye.) However, this happens rarely—usually the planet is a little above the solar disk, or a little below. All the usual warnings about looking into the sun apply.

Most reports of naked-eye transits of Venus (or, even less believably, Mercury) likely are reports of sunspots instead. The great Islamic scholars Al-Kindi (circa 800–70), Ibn Rushd (Averroes; 1126–98), and Ibn Bajja

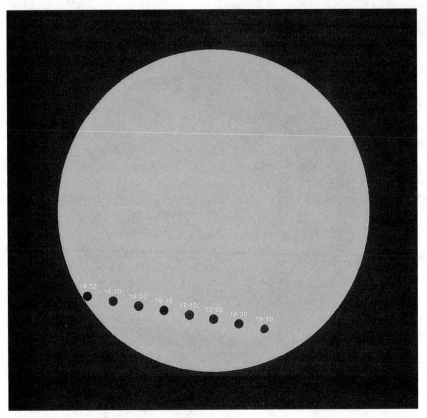

FIGURE 12.3.   A time-lapse image of Venus (silhouette) transiting the sun, June 8, 2004. This is a telescopic image; without optical aid, Venus would be, at most, a dark speck. Photo by Shigemi Numazawa.

(Avempace; died 1138) were all thusly fooled. So was the founder of modern planetary astronomy, Johannes Kepler. When we read in history of such an event lasting for days, it cannot be caused by something revolving about the sun; that is too long. It must be due to a phenomenon on the disk of the sun itself.

Transits of Venus occur in pairs every 128 years, with eight years in between. The most recent pair is 2004 and 2012. Venus transits are always in the months of December or June.

Planets shine only by reflecting sunlight. The brightness of a planet in our sky is a function of its distance from the sun, its distance from the earth, and the angle of illumination. It is also a function of the planet's albedo and its apparent size as seen from the earth. Even though we cannot see the disk of a planet with the naked eye as we can the moon, it is true for both the moon and planets that the bigger the apparent size of its disk, the brighter.

Based on the above, you might think that an inferior planet would be brightest around inferior conjunction. However, just like the moon, inferior planets undergo phases. Venus, for example, is a crescent as seen—telescopically—from the earth near inferior conjunction (a geometry analogous to new moon).

Can anybody see the crescent of Venus with the naked eye? It is not thought possible. (Look at figure 9.3 again; it was photographed with a camera/telescope superior to the naked eye.) Still, there are ancient Mesopotamian personifications of Venus that associate her with a crescent symbol. Was it a lucky guess?

If Venus is a crescent, not much of its disk is in sunlight; its brightness averaged over the area of the disk is diminished. But at "full Venus," the planet is far from us, at superior conjunction, and presents to us a disk of smaller apparent size than it does at crescent phase. This shrinkage reduces the planet's brightness, too. These competing factors result in Venus's greatest brightness in our sky as a crescent near maximum elongation. Venus is most brilliant about thirty-six days before and after inferior conjunction.

At its brightest, Venus is fifteen times brighter than the brightest star on the celestial sphere, Sirius. It can be seen at any air mass—right to the horizon. In a dark, dark setting, Venus can cast shadows. (Jupiter can, too.) As a planet always near the horizon and changing in appearance quickly, from day to day, no wonder that many UFO reports are attributable to Venus.

The light from Venus reflects off uniform, white clouds in that planet's atmosphere. The light from Mercury reflects off that atmosphere-less world's surface. The surface of the planet Mercury is similar to that of the moon, and its brightness as a function of elongation is similar to the moon's brightness as a function of phase. The result is a more simple brightening behavior than that of Venus. As you might intuit, Mercury is brightest several days before and after "full Mercury" (Mercury's superior conjunction).

SUPERIOR PLANETS DON'T FEEL THAT WAY

The planets that are farther from the sun than the earth are superior planets. There are more of these: Mars, Jupiter, and Saturn in that order from the sun. (An exceptionally good eye might be able to detect Uranus or one of the minor planets (asteroids), under the right conditions; still, we will ignore these.) With superior planets we do not get the same geometries we got with the inferior planets. Because the orbit of a superior planet encompasses both the sun and the orbit of the earth, a superior planet is never between us and the sun. It would have to jump out of its orbit to become so.

Superior planets have only one conjunction. While the planet can be behind the sun as seen from the earth, it never can get in front of the sun as seen from the earth. We do not have to distinguish between superior or inferior conjunction—it is just conjunction, when the planet is as far from the earth as possible. A superior planet must reach conjunction once every synodic revolution.

Yet, notice that a superior planet also can show up where no inferior planet can: For instance, a superior planet can be sixty degrees away from the sun, a point called sextile; ninety degrees away, a point called quadrature; or 120 degrees away, trine. In fact, it can have any elongation.

Instead of reaching a second conjunction, a superior planet can line up with the earth and sun on the opposite side of the earth from the sun. At this point, it is also true that the planet is as close as it ever gets to the earth.

Because of this proximity and the fact that the superior planet is at this time up all night, high in the sky at midnight (as far from the sun as something can be in our sky), here it is most easily observed from the earth. The position is called opposition. A superior planet also must reach opposition once every synodic revolution.

The time during which a superior planet is visible to us (not near conjunction) is called an apparition. Opposition occurs halfway through each apparition.

In chapter 7 you learned about the heliacal rising of a star. A planet can rise heliacally as well. What is more, a superior planet can rise acronically. (It can set acronically, too, for that matter.) An inferior planet cannot. These not-well-defined acronical events, for stars or planets, do not seem to have made the mark in history that heliacal ones have, though.

While superior planets appear to travel in a prograde direction most of the time, near opposition they undergo a brief period of retrograde motion, before returning to prograde. The points of transition from prograde to retrograde motion, and from retrograde to prograde motion, are called a planet's stationary points.

The so-called retrograde loop is confusing and unexpected, but explainable by the fact that superior planets orbit the sun more slowly than does the earth. (Saturn is slower than Jupiter, which is slower than Mars.) As opposition approaches, the earth overtakes the superior planet. Thus the superior planet appears, for a time, to be moving backward as viewed

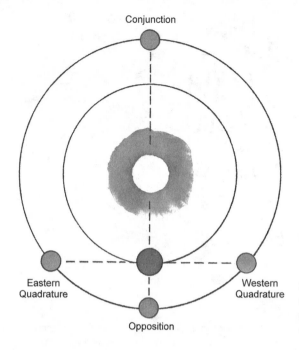

Conjunction

Eastern
Quadrature

Western
Quadrature

Opposition

FIGURE 12.4. A superior
planet in four different
geometries with respect to the
earth and sun. Not to scale.

against a distant background (the stars). You see the same thing from a
car passing another car, going into a curve on the highway.

Notice that we call it a "loop," not a "zigzag." The "loop" part of retrograde motion comes from the fact that, even if a planet pauses in its
eastward or westward motion, it is still traveling a little north or south:
The planes of the planetary orbits are all inclined slightly to the ecliptic.

Mars demonstrates the most profound retrograde loop. This is because it is the closest planet to us that is moving in its orbit more slowly
than we are. (Its synodic period is 780 days.) The more-distant planets
move in smaller retrograde loops.

The brightness of a superior planet also is governed by how big its
disk appears at the earth just then, its albedo, the angle of illumination,
how far the planet is from the source of sunlight, and, of course, how far
away the planet is from us. (Because a superior planet is always backlit, it
does not exhibit phases the way an inferior planet does; at most, it may
appear a bit gibbous through a telescope.) That last factor is minimized
at opposition. The planet is then at its brightest for that synodic period,
though variations in the other factors can affect its absolute brightness
at any time during a given synodic revolution.

In the case of Mars, whether the earth or Mars is near perihelion or
aphelion can noticeably affect Mars's brightness at opposition. Martian

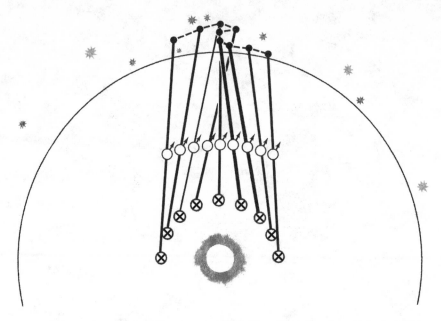

FIGURE 12.5. An explanation of the retrograde motion of Mars. The circle-and-arrow is the symbol for Mars. The symbol for the earth is the circumscribed "X." Each straight line represents the line of sight from the earth at a different time. Not to scale.

oppositions are best when they occur in August. Such favorable oppositions of Mars happen every fifteen or seventeen years.

Even then, some favorable oppositions are more favorable than others. For reasons not entirely understood by planetary geologists, one hemisphere of Mars is darker than the other. Mars has an obliquity, just like the earth. Which hemisphere is pointed toward us at opposition causes Mars's reflectivity to vary and, with it, Mars's brightness in our sky.

At no time does Mars become close enough to the earth so that we can see its disk with the naked eye. Each year, the inevitable Internet "viral" makes its way into my in-box. I refer to the one that states that *this* year (whichever year it is) Mars will be closer to the earth than it has been in some number of years, and that Mars will appear as a disk the size of the full moon. Sometimes it is exactly the same e-mail as the year before. Naturally, no source is ever cited for this astonishing information. Like every other such anonymous e-mail I have ever received, it is false.

Jupiter is normally the brightest of the superior planets in our sky, rivaling Venus. Jupiter is not nearly as close to the earth as are Venus or Mars, but it is very large and has a high albedo.

The explanation of a variation in the brightness of Saturn at opposition had to await the telescopic discovery of that planet's rings. The rings orbit Saturn's equator. Saturn's obliquity causes the rings to be tilted periodically so that sometimes they present their surfaces obliquely toward us, while at other times they are edge-on as seen from the earth. The rings of Saturn are shiny, and the unresolved planet-plus-rings reflects more light toward us every fifteen years when the rings are tipped in our direction and their surface area, from our perspective, is maximized.

PLANET	AVERAGE SYNODIC PERIOD (DAYS)	TYPICAL DURATION OF RETROGRADE MOTION (DAYS)	TYPICAL INVISIBILITY DUE TO APPARENT PROXIMITY TO THE SUN (DAYS)[3]
Mercury	116	22	5 and 35
Venus	584	42	8 and 50
Mars	780	72	120
Jupiter	399	120	32
Saturn	378	138	25

PLANET MEETS PLANET

Planets come into conjunction with the sun; planets can come into conjunction with other planets, too. Any time one planet is near another in our sky, it is commonly called a conjunction. Technically, a conjunction implies that the two planets cross the celestial meridian at the same time—that is, they are angularly close, and north and south of each other.

For centuries, people have tried to equate the so-called Star of Bethlehem with a historical conjunction of planets—in what may be the symbolically Christian constellation of Pisces. Today, many scholars agree that the conjunction of 7 BCE (the most likely candidate) was too early to signal the imminent birth of Jesus. Another candidate is the conjunction of Venus and Jupiter in 2 BCE. But would respectable magi see a female symbol (Venus) as a signal for a male king's birth?[4] And to be fair to the Gospel, a casual reading of Saint Matthew might imply a *miraculous* Star of Wonder, instead of a pedestrian natural event such as a planetary conjunction.

Three or more planets in close proximity on the celestial sphere are called a massing. However, the grouping of all three naked-eye superior planets is given a special name, a grand conjunction. Grand conjunctions occur only about every twenty years or so. The last was in the year 2000. (It was an unusual event because Saturn, Jupiter, and Mars were joined by Venus and Mercury in a five-element massing.)

It is the periods of Jupiter and Saturn that govern the frequency of grand conjunctions. These planets move so slowly on the celestial sphere, compared to Mars, that Mars always can be counted upon to catch up with them sometime during a Jupiter-Saturn conjunction.

Harvard University's Owen Gingerich reminds us that Nicolaus Copernicus (1473–1543) watched Saturn, Jupiter, and Mars approach each other in 1503/1504. Doing so, he may have wondered whether there might be a better way to model the solar system than the earth-centered view of his predecessors.

The ancient Greeks went to extremes. They imagined a time, in the remote past or future, when all the planets were tightly massed (or in some other interesting configuration). They recognized that the time it would take for all the planets to repeat exactly the same pattern in the sky would be fantastically long. Still, they gave a name to this theoretical interval, the Great Year. After the Great Year, all was expected to repeat. The idea of cyclic repetitions of history appeared in the New World, too, as we will see below.

An actual occultation of a bright star by a planet is uncommon. The most recent one was the occultation of Regulus by Venus in 1959.

On the other hand, the planets (as well as the sun and moon) routinely cross the Milky Way and/or one of the rifts within it. While not an occultation in the formal sense, such an event may have been of interest to indigenous people, particularly of the Southern Hemisphere.

As you might imagine, an occultation of a planet by a planet is rare. History has recorded two: Mercury by Venus in 1737 and Jupiter by Mars in 1591.

PEOPLE AND PLANETS

The planets neither provide significant warmth nor illumination for human needs. Why keep track of them?

When we think of the ancient Maya, an Indiana Jones landscape of vine-draped ruins comes to mind. But before the jungle claimed them, what are now ruins were once vibrant cities of stone, sheltering a people seemingly obsessed with cycles in the sky. We know this because a precious few surviving Maya books deal with calendrical cycles, the most famous being the *Dresden Codex*.

Why? The Maya felt that the past was prologue. They subscribed to a worldview that saw history as repeating. They looked for and found recognizable patterns of repetition in nature. The Maya were well versed in both the tropical year and the synodic month by the sixth century BCE. They also kept a 260-day calendar, the basis for which is debated to this day. (Regardless of how it evolved, a 260-day agricultural period is useful in the climate of the Yucatán.) Most amazing, to me, is that the Maya kept track of the morning/evening star, Venus, as well. Not only did they determine its 584-day synodic period, they cleverly noticed that 584 days is commensurate with 365 days (the year) once every eight years. (This is also ninety-nine synodic months, an interval named the octaeteris.)

They did all this for good reason. To the Maya, Venus was a god. Plotting its motions in the sky was akin to consulting a deity about the decisions of the day.

The Maya civilization fell into eclipse at about the time of the European invasion of Mesoamerica. We might know more about their astronomy had less of their writings been burned by overzealous Christian missionaries.

A superior planet was of great import to the Chinese. We all have visited a Chinese restaurant where, on the placemat, we can look up when we were born in the Chinese birth-year series. Each year corresponds to an animal (and, supposedly, to certain character traits). The cycle is twelve-years long; a "rat" is born twelve years after another "rat."

Where did the twelve come from? Answer: the sidereal period of Jupiter, often the brightest planet in the sky. (Only Venus can top it in absolute brightness under certain circumstances.) Each year, Jupiter moves into another one of the twelve Chinese houses, or *tzhu*. Jupiter is *sui hsing*, the Chinese Year Star. The sidereal period of Jupiter is not exactly twelve years, though. About every eighty-four years, the Chinese chronologists have to fudge the calendar by one house in order to keep on track.

The planets play a big role in Western astrology. A phrase like "when Jupiter aligns with Mars" is so familiar that it has become a cliché.

The Maya Venus calendar, the Chinese Year Star, Western astrological practices—none of these have any utilitarian use. But since when are our lives consumed totally with day-to-day practical matters? What is more, if you believed that planets such as Venus or Jupiter *did* foretell future events—in an uncertain world, they were practical indeed.

PLANETS VERSUS STARS

Clearly, we cannot be expected to mentally keep track of the complicated motions of the planets, like we might keep up with the phases of the moon or the seasons. A chart, table, or map is gratefully consulted. There is even a special term for published daily positions of the planets—ephemerides. But if one walks outside without such aids, is it possible to find the planets on one's own?

Notwithstanding some peoples' claims to be able to see the phases of Venus, the rings of Saturn, or the satellites of Jupiter with their naked eye, to most of us, planets look suspiciously like stars. Which of the points in our star-spangled sky are not stars but planets?

Except for Mercury, the planets are fairly bright. Mars is noticeably red, Jupiter is yellowish. All this helps. However, there are bright, colored stars in the sky, too.

Better—by memorizing the major stars in the zodiac, you stand a good chance of picking out a star that "doesn't belong there." A few nights later you may be able to confirm your planet find by noting its motion relative to the fixed stars—at least for the faster moving planets such as Mercury, Venus, and Mars.

Or try this trick: Look for twinkling. The song says "Twinkle, Twinkle Little Star," and for a reason. Planets do not twinkle as much as stars. Twinkling is produced by starlight passing through the earth's undulating atmosphere. Stars are point-like. Because planets are actually (tiny) disks, just too small to be resolved by our unaided eye, their light does not act as if it is coming from a single point. On a night in which the atmosphere is fairly stable, they are less susceptible to the twinkling effect.

An exception is Mercury. We are normally forced to view Mercury low in the sky (and, thus, through high air mass). In addition, it is usually seen when in the crescent, more point-like phase. So Mercury almost always is twinkling—just like a star. Indeed, on certain nights of unstable air, everything in the sky twinkles.

WE ALL HAVE DAYS

It probably has occurred to you that the names for the days of the week have something to do with the sky. Sunday sounds a lot like "Sun Day" and Monday sounds much like "Moon Day." You are right. The seven days of the week correspond to the seven "planets" visible in the ancient sky: the sun, moon, and five objects that today we call the naked-eye planets.

This identification is obscured in English by the fact that our days of the week refer to the mythological gods associated with the planets. While we still use the Roman names for the planets, we insert the Germanic counterparts of these gods when we refer to days. For instance, the war god in Old English, named Tyr, was a natural stand-in for the Roman Mars. Tuesday is "Mars' Day." OK. Maybe you have not heard of Tyr, but I bet you know Thor (if only from a comic book). Thor was matched up with the chief Roman deity Jupiter. Thursday equals "Jupiter's Day." Frigg was a goddess, so she became the only female of the bunch—Venus. Friday is "Venus's Day." Who was left? Wodan became Mercury. Wednesday is "Mercury's Day." Saturn was left alone—Saturday is "Saturn's Day."

So far, so good. But what about the sequence for the days of the week? A tradition that goes back to the Greeks holds that the order of the "planets," in decreasing distance from the earth, is Saturn, Jupiter, Mars, sun, Venus, Mercury, and moon. (They actually got Saturn, Jupiter, and Mars correct, and the moon is indeed the closest of these bodies to the earth.) Yet that is not the order of our days.

Here the astrologers stepped in. They decided that every hour of the day was governed by a "planet." If the day began with Saturn, each of the subsequent twenty-three hours of that day would be governed by another planet in the Greek order:

> Hour 1 = Saturn
> Hour 2 = Jupiter
> Hour 3 = Mars
> Hour 4 = Sun
> Hour 5 = Venus
> Hour 6 = Mercury
> Hour 7 = Moon

Now we are out of "planets." So we must return to the top of the list.

> Hour 8 = Saturn
> Hour 9 = Jupiter
>
> . . . .
>
> Hour 14 = Moon

And again and again:

> Hour 15 = Saturn

All the way to:

Hour 24 = Mars

Thus, the first hour of the new day is

Hour 1 = Sun

Eventually each day in its entirety was tagged with the planet associated with the first hour of that day. If you go through the list a few more times, you arrive at the moon for the first hour of the third day. And Mars for the first hour of the fourth day. And so on. The result of all of this is: Saturday, Sunday, Monday, Tuesday, Wednesday, Thursday, and Friday. (Christianity insists that the week begins with Sunday, but other religions beg to differ.)

FLEETING PHENOMENA OF THE NAKED-EYE SOLAR SYSTEM

There are a few more celestial oddities that you may see in the sky. While everybody likely has seen stars, the sun, the moon, and (perhaps unknowingly) a planet, these are more "catch as catch can."

Beyond most of the planets is the home of comets, irregular lumps of ice and dirt. That does not sound very aesthetic, but, in fact, comets can be lovely. When occasionally their long, elliptical orbits bring them closer to the sun (and nearer us), the sun's warmth vaporizes the volatile ice. The small, solid comet nucleus becomes surrounded by a many-powers-of-ten-bigger envelope of bright gas, which then flows away from the sun. The resulting shiny, fuzzy ball and tail can be the most beautiful sight in the sky.

All of this may become bright enough to see with the unaided eye. Most of the time, though, the particular comet was unknown before it visited our solar-system neighborhood. In other words, the appearance of bright comets is often unpredictable months or even weeks ahead of time.

This is a book about generally observable phenomena in the sky. So I am going to resist the temptation to write much more about comets here. However, I wrote an entire book just about these visitors, called the *Comet Hale-Bopp Book* (1996). It was written (quickly!) before the best appearance of an especially big, newly discovered comet named Hale-Bopp. Still, most of the information applies to future comets, too. If you hear of a newly spotted comet heading our way, or simply want to know more about these unique inhabitants of our solar system, check it out.

Some people think of a comet as something that flies by so fast that

Greg Bock, Comet McNaught P1 2006, from Leyburn, 20070120.
Copyright, Bannockburn Observatory, Queensland Australia

FIGURE 12.6. Comet McNaught was seen best from the Southern Hemisphere. Here it is photographed from Leyburn, Queensland, Australia on January 20, 2007. Photo by Greg Bock. Canon 350D camera, ninety-six-second exposure at ISO400.

they may miss it if they blink at the wrong time. No, comets may grace our sky for many nights or weeks of nights. These people likely are confusing comets with meteors.

On a much smaller scale, now, there is a myriad of small solar-system bodies between the planets. These objects revolve about the sun in random orbits, and normally are invisible to us. These small bits of rock and metal are called meteoroids.

Meteoroids are leftover pieces from the formation of the solar system, the ejecta of impact craters, or the result of asteroids bumping into each other (chips off the "old block"). Some clusters of meteoroids are thought to be the spent remains of comets.

Meteoroids are what cause craters, when they strike the solid surfaces of larger bodies (like the moon).

On the earth, though, most are destroyed by the extreme frictional heat of these fast projectiles tearing through our atmosphere on their way to the ground. We can see their fiery deaths: They are called "shooting stars," "falling stars," or, more properly (because they have absolutely nothing to do with stars), meteors.

Meteors are not technically part of the celestial realm, the purview of this book. Still they do appear in the sky, randomly and, more or less, continuously. Human beings spend little time looking up at the sky; therefore, when one does happen to glance up and catch sight of a meteor, it is considered by some to be a sign of good luck.

Imagine: You spend hundreds of millions of years traveling around the solar system, minding your own business; then, suddenly one day, up ahead, there is this big blue ball in the way. And in not much more than an instant, you end up as a bright streak and bit of ash that somebody makes a wish upon.

The typical meteoroid I have been discussing weighs less than a gram and is the size of a dust grain. Sometimes, though, a larger meteoroid actually does make it to the ground—or what is left of it does. Because it has been slowed by passage through the atmosphere, it may even land relatively softly. Someone then can come along and pick it up like a rock. So it is given a new name that sounds like a rock: Now it is a meteorite.

"Meteoroid," "meteor," and "meteorite." Be careful of these three words—they sound a lot alike. Notice that a meteor is not so much a thing as a visible phenomenon caused by a thing. A meteor is a flash of light in the sky. Like lightning, it is not a solid object. However, a meteor is caused by a solid object (the heated meteoroid).

Sometimes the earth intercepts a whole swarm of meteors. Then, the number of meteors you can see in the sky on a given night gets a little bit higher and we have a meteor shower. As the earth arrives at the same place in its orbit at the same time each year (the place where the swarm is, say), meteor showers occur annually.

Such a meteor shower is misnamed. The frequency of meteors during these days or weeks usually is not as great as the word "shower" implies. (Maybe "meteor sprinkling" would be better.) Nevertheless, there have been spectacular exceptions when meteors literally stormed downward at uncountable rates.

At one time, meteors were believed to be a strange weather phenomenon and to have nothing to do with bodies in space. Then it was noticed that meteors in a given shower all seem to come from the same point on the celestial sphere. Each streak diverges from this point, called the shower radiant. The effect is the same if one watches a line of cars, side by side, traveling toward you on the highway. At the horizon, the cars all seem to occupy a single point, but as they get closer, they appear to move outward from this point. (When this happens, it would be a good idea to stop standing in the middle of the road.) A radiant indicates that meteoroids originate at a great distance beyond the earth's atmosphere.

FIGURE 12.7. A bolide, September 30, 2008, seen in western Oklahoma. Photo by Howard Edin (Oklahoma City Astronomy Club).

Shower meteoroids rarely result in a meteorite. Those meteoroids are bigger and random. They produce sporadic meteors. A larger object may put on a spectacular display as it enters the atmosphere. Instead of a mere streak of white light, we may be treated to a varying blur of flame that last several seconds, changes color, and leaves behind a trail of smoke. Its path even may change, like a pitcher's curveball. We call these infrequent fireballs bolides. Rarely, bolides may be visible in daylight. In modern times, it is possible that a bolide might be the demise of manufactured space junk.

On the night of December 5, 2008, a bolide the brightness of one hundred full moons dropped in on the state of Colorado. It appeared directly above a camera pointing upward—for the expressed purpose of photographing meteors. How is that for "wishing star" luck?

Want to increase your odds of spotting a meteor? During a meteor shower, look when the constellation in which the shower radiant resides (and after which the shower usually is named) is above the horizon. On other nights, morning hours (before twilight) tend to yield more meteors than do late evening hours (after twilight). This is because, at these morning hours, our sky is facing into the path of the earth about the sun.

Earlier, we have the earth between us and where most of the action is taking place.

The inner solar system is dusty. Micrometeoroids cannot be seen individually without a microscope. But in quantity they do make their presence known. Under optimum dark conditions, it is possible to see a pyramid of faint light towering into the sky away from the just-set or the soon-to-rise sun. This zodiacal light follows the ecliptic and is slightly brighter 180 degrees from the sun (where it is then called gegenschein, a German word for "counterglow"). Look for the zodiacal light in the spring and fall, when the ecliptic is more perpendicular to the horizon. The zodiacal light is interplanetary dust faintly illuminated by the sun. For most of us, it is also the celestial phenomenon that marks the limit of our naked-eye vision.

STUFF YOUR GREAT-GRANDPARENTS NEVER SAW

The space age has added objects to our naked-eye sky. It is comparatively easy to see artificial earth-orbiting satellites moving between the stars. They travel much more rapidly than the moon or planets do, yet are not as quick as meteors. A satellite usually takes minutes (not hours or seconds) to cross the sky.

Do not confuse satellites with aircraft. If what you are watching consists of colored lights, has multiple lights, or exhibits blinking lights, it is not a satellite. Satellites are so far away that they cannot be resolved by the unaided eye. They are point-like, moving "stars." If the satellite is tumbling, it may change in brightness regularly.

Look for satellites in the early evening or still-darkened morning. Even then, they will appear and disappear while still some angular distance from the horizon. Satellites are close enough to the earth that at any lower altitude they are submerged in the earth's shadow. We cannot see any light that might be emitted by the satellite—it is too faint. Like all other solar-system objects sans the sun, a satellite shines only by reflected sunlight. As the hour approaches midnight, more and more of a satellite's above-horizon path is in the earth's shadow. After midnight, less and less.

Is the satellite traveling north to south? Or vice versa? Polar orbiting satellites are more expensive, but with the earth rotating underneath them, they can "see" every longitude and latitude, eventually. (More equatorially orbiting satellites never have a good view of high latitudes.) These satellites often are used for reconnaissance. So if you spot one, smile for the camera.

FIGURE 12.8. Zodiacal light from the Big Island of Hawaii. Photo by Shigemi Numazawa.

The biggest artificial satellite is guaranteed to appear bright if it is in sunlight while it passes near you. This is the International Space Station (ISS). It is made brighter from time to time by a visiting space shuttle or Russian Soyuz vehicle. My son and I fondly remember watching the ISS and shuttle flying in formation, near to each other before docking—a "double star" of lights dancing together across the sky. As you watch such crewed spacecraft, imagine somebody aboard looking back at you.

A favorite satellite of mine is the Hubble Space Telescope. We are in the same business at the same time: sky watching.

The visibility of all these satellites, for your location, can be known in advance. To find predicted times and directions in which to look, enter "satellites," or the proper name of a satellite, into a World Wide Web browser.

One series of satellites deserves special mention. They are the Iridium satellites, used for telephone communication. Parts of the Iridium satellites are especially shiny—mirrorlike, in fact. If the glint of the sun strikes them and reflects back to you, the result is amazing: For a second or two, the satellite is far brighter than any star or planet. It may rival the moon. Besides the position of the satellite, the sun, and you, these Iridium flares are governed also by the orientation of the satellite. Yet they, too, can be predicted.

When I take students outside and point out the North Star, they often are disappointed. By reputation, most people expect it to be a much brighter star than it is. No one, though, was disappointed on one particular evening: As usual, I aimed my green laser at Polaris. As I did so, a brilliant star immediately burst forth, brighter than anything else in the heavens. My pointer, now transformed into a magic wand, had conjured up a sight worthy of the awe reflected in my students' faces. A totally coincidental Iridium flare, appearing in the constellation of Ursa Minor, had made me the Wizard of the Evening.

I hope that I have convinced you that the sky is a busy place. This is one of many reasons why I am unimpressed by those who would further populate it with exotic objects such as alien spaceships. Notwithstanding the logical objections to such fancies, there is a much simpler explanation for those who think that they have been chased by a "flying saucer."

Because astronomical objects are so far away, while you may be moving on the earth (say, driving a car), the angular motion with respect to an astronomical object is minuscule. An example: If you are driving north and see Venus (out of your left-hand window) to the west, it will stay there until its own motion in the sky becomes appreciable. But this takes many minutes. Meanwhile, your sensation is that you are moving quite fast. In our brains' normal experience, this scenario only happens when the object out our window is moving at the same rate we are (like another car in the next lane might). If you usually are not aware of the night sky, it is easy to be surprised by a bright planet and convince yourself that it is a much closer object, nature unknown, pursuing you.

The planets do not pursue us. However, the show they put on in the sky never repeats itself in a lifetime. Planets are worth pursuing.

NOTES

INTRODUCTION

1. The Beatles, "Magical Mystery Tour," Apple, 1967.

CHAPTER 1

1. To "visit" archaeological sites mentioned in this book, see the accompanying website, www.press.uchicago.edu/books/hockey.

2. J. McK. Malville, R. Schild, F. Wendorf, and R. Brenner, "Astronomy of Nabta Playa," *African Skies* 11 (2007): 2. (Where I insert a footnote to provide the reader with more detail, I normally will try to choose a source that is readily accessible.)

3. Philip C. Steffey, "The Truth about Star Colors," *Sky & Telescope* 84, no. 3 (1992): 266.

4. Nicholas T. Bobrovnikoff, *Astronomy before the Telescope, Volume I: The Earth-Moon System* (Tucson, AZ: Packart Publishing House, 1984).

5. Raymond Haynes, Roslynn Haynes, David Malin, and Richard McGee. *Explorers of the Southern Sky: A History of Australian Astronomy* (Cambridge: Cambridge University Press, 1996).

6. Alan MacRobert, "The Hue of the Universe," *Sky & Telescope* 103, no. 4 (2002): 21.

CHAPTER 2

1. Quoted by Damian U. Opata, "Cultural Astronomy in the Lore and Literature of Africa," in *African Cultural Astronomy: Current Archaeoastronomy and Ethnoastronomy Research in Africa*, ed. Jarita Holbrook, Rodney Medupe, and Johnson Urama (New York: Springer, 2008).

2. To see astronomical instruments mentioned in this book, visit the accompanying website, www.press.uchicago.edu/books/hockey.

3. George Lovi, "Directions—Earth and Sky," *Sky & Telescope* 77, no. 4 (1989): 399.

4. Astronomical Society of the Pacific Conference Series 89 (1996): 291.

CHAPTER 3

1. Germano B. Afonso, Maria C. Beltrão, and Adnir A. Ramos, "Archaeoastronomy in Florianópolis (Brazil)," in *51° Congreso Internacional de Americanistas,* ed. Johanna Broda, Gonzalo Pereira, and Maxime Boccas (Santiago: Universidad de Chile, 2003).

2. Instead of lists of people who have applied a particular astronomical concept in their culture, I usually will pick out one example. I will choose such examples to span great breadths of geography and great spans of time, the purpose being to suggest that astronomy-in-culture is global and current in all human epochs.

CHAPTER 4

1. Many of the star names in use today come to us from the Arabic language.

2. John Isles, "The Top 12 Naked-Eye Variable Stars," Sky & Telescope 93, no. 5 (1997): 80.

3. F. Richard Stephenson, "Early Chinese Observations and Modern Astronomy," Sky & Telescope 97, no. 2 (1999): 48.

4. Some historians argue that abbreviations such as "BC" and "AD" are inappropriate, because their root meanings lack significance in many parts of the world. These scholars use abbreviations such as "CE" and "BCE," which stand for "Common Era" and "Before the Common Era," respectively. (The year 1 BCE is exactly the same as the year 1 BC, the year 1 CE is exactly the same as the year 1 AD.) Others argue that "AD" and "BC" have been in use for so long that to change now is confusing. I have no strong opinion on this issue either way!

5. David W. Hughes and Carole Stott, "The Planisphere: A Brief Historical Review," Journal of the British Astronomical Association 105 (1995): 35.

6. William C. Miller, "Dark Adaptation: Its Nature and Preservation," AAS Photo-Bulletin 24 (1980): 18.

7. Keith Bowen, "Aging Eyeballs," Sky & Telescope 82, no. 3 (1991): 254.

CHAPTER 5

1. Greek for Sirius

2. For more on this, see Anthony Aveni, Conversing with the Planets (New York: Kodansha International, 1994).

CHAPTER 6

1. Popular Astronomy 44 (1936): 135.

2. There is no popular holiday observed at this time of year within the entire United States. However, thirteen states have declared a sales-tax-free holiday on or near the first weekend in August. A cross-quarter day observance? Yes and no. This weekend also is near the beginning of the traditional school year.

CHAPTER 7

1. The word "arctic" comes from the Latin word for "bear." The prominent arctic constellations of Ursa Major and Ursa Minor are, in Latin, the Great Bear and Lesser Bear, respectively.

2. Author Anne Rice was confused, too. In her screenplay for Interview with the Vampire (1994; based on Rice's book), two "undead" characters are trapped at the bottom of a cistern with the sun shining down upon them. (We learn that this is really bad for vampire health.) But, no: that misfortune could not happen within European latitudes. One less thing vampires have to worry about.

3. Clive Ruggles, "Astronomy, Oral Literature, and Landscape in Ancient Hawai'i," Archaeoastronomy: The Journal of Astronomy in Culture 14, no. 2 (1999).

4. Sixto Ramón Giménez Benítez, Alejandro Martín López, and Anahi Granada, "The Sun

and the Moon as Marks of Time and Space among the Mocoví of the Argentinean Chaco," *Archaeoastronomy: The Journal of Astronomy in Culture* 20 (2006): 54.

5. Daryn Lehoux, *Astronomy, Weather, and Calendars in the Ancient World: Parapegmata and Related Texts in Classical and Near-Eastern Societies* (Cambridge: Cambridge University Press, 2007), 51.

CHAPTER 8

1. Written by Graeme Edge and appearing on the album *Days of Future Past* (1967).

2. "As all the apes ... show a longer menstrual period than that of modern humans, the last common ancestor of chimpanzees and humans is likely to have had a longer menstrual period than modern humans, too. . . . According to this evolutionary trend, the menstrual period of the human lineage would be expected to undergo further shortening in the future . . . no temporal relationship emerges between the duration of the lunar month . . . and the menses." M. Folin and M. Rizzotti, "Lunation and Primate Menses," *Earth, Moon, and Planets* 85–86 (2001): 539.

3. Clive Ruggles, "Four Approaches to the Borana Calendar," in *Archaeoastronomy in the 1990s*, ed. Clive Ruggles (Loughborough, UK: Group D Publications, 1993).

4. Dennis Tedlock, and Barbara Tedlock, "Moon Woman Meets the Stars: A New Reading of the Lunar Almanacs in the Dresden Codex," in *Skywatching in the Ancient World: New Perspectives in Cultural Astronomy,* ed. Clive Ruggles and Gary Urton (Boulder: University Press of Colorado, 2007).

5. Martha J. Macri, "A Lunar Origin for the Mesoamerican Calendars of 20, 13, 9, and 7 Days," in *Current Studies in Archaeoastronomy: Conversations across Time and Space,* ed. John W. Fountain and Rolf M. Sinclair (Durham, NC: Carolina Academic Press, 2005).

6. E. C. Krupp, "Cows Bound for the Moon," *Sky & Telescope* 90, no. 6 (1995): 60.

7. Roy E. Hoffman, "Back-to-Back Crescent Moons," *Observatory* 129, no. 1208 (2009): 1.

8. K. D. Abhyankar, "Synodic Month in Hindu Pañchānga," *Observatory* 111 (1991): 315.

9. Linda Amy Kimball, "The Batak Porhalaan Traditional Calendar of Sumatra," *Archaeoastronomy* 11 (1989–93): 28.

CHAPTER 9

1. D. McNally, "The First 400 Years of the Gregorian Calendar," *Irish Astronomical Journal* 16 (1983): 17.

2. Nathan Sivin, *Granting the Seasons: The Chinese Astronomical Reform of 1280, with a Study of Its Many Dimensions and an Annotated Translation of Its Records* (New York: Springer, 2009).

3. Philip Hiscock, "Folklore of the Blue Moon," *The Planetarian* 22, no. 4 (1993).

CHAPTER 10

1. Kevin Krisciunas, "Determining the Eccentricity of the Moon's Orbit without a Telescope," *American Journal of Physics* 78 (2010): 828–33.

2. W. G. Rees, "The Moon Illusion," *Quarterly Journal of the Royal Astronomical Society* 27 (1986): 20.

CHAPTER 11

1. "Totality's End," *Sky & Telescope* 111 (2006): 96.

2. F. Richard Stephenson, "Investigation of Medieval European Records of Solar Eclipses," *Journal for the History of Astronomy* 41, pt. 1 (2010): 95.

3. Breading G. Way, "An Unusual Eclipse Year," *Publications of the Astronomical Society of the Pacific* 46 (1934): 346.

4. Tony Freeth, Alexander Jones, John M. Steele, and Yanis Bitsakis, "Calendars with Olympiad Display and Eclipse Prediction on the Antikytherea Mechanism," *Nature* 454 (2008): 614.

5. Emília Pásztor, "Some Remarks on the Moon Cult of Teutonic Tribes," in *Archaeoastronomy in the 1990s*, ed. Clive Ruggles (Loughborough, UK: Group D Publications, 1993).

CHAPTER 12

1. Only Mercury's orbit is inclined significantly to the ecliptic.

2. Mercury's orbit is particularly eccentric. Maximum elongation may occur at an even smaller angle if Mercury also at the time happens to be close to perihelion.

3. Table data from M. Zeilik, *Astronomy: The Evolving Universe*, 9th ed. (Cambridge: Cambridge University Press, 2002), and S. Milbrath, *Star Gods of the Maya: Astronomy in Art, Folklore, and Calendars* (Austin: University of Texas Press, 1999).

4. Richard Coates, "A Linguist's Angle on the Star of Bethlehem," *Astronomy & Geophysics* 49, no. 5 (2008): 27.

RECOMMENDED READING

Aveni, Anthony F. *Skywatchers*. Austin: University of Texas Press, 2001.

Bobrovnikoff, Nicholas T. *Astronomy before the Telescope*. Vols. 1–3. Tucson, AZ: Pachart Publishing House, 1984–90.

Brody, Judit. *The Enigma of Sunspots: A Story of Discovery and Scientific Revolution*. Edinburgh: Floris Books, 2002.

Harrison, Edward. *Darkness at Night: A Riddle of the Universe*. Cambridge, MA: Harvard University Press, 1987.

Heilbron, J. L. *The Sun in the Church: Cathedrals as Solar Observatories*. Cambridge, MA: Harvard University Press, 1999.

Kaler, James B. *The Ever-Changing Sky: A Guide to the Celestial Sphere*. New York: Cambridge University Press, 1996.

Milbrath, Susan. *Star Gods of the Maya: Astronomy in Art, Folklore, and Calendars*. Austin: University of Texas Press, 1999.

Montgomery, Scott L. *The Moon in the Western Imagination*. Tucson: University of Arizona Press, 1999.

Neugebauer, Otto. *The Exact Sciences in Antiquity*. Mineola, NY: Dover Publications, 1969.

Reed, George. *Dark Sky Legacy: Astronomy's Impact on the History of Culture*. Buffalo, NY: Prometheus Books, 1989.

Ruggles, Clive. *Ancient Astronomy: An Encyclopedia of Cosmologies and Myth*. Santa Barbara, CA: ABC Clio, 2005.

Schafer, Edward H. *Pacing the Void: T'ang Approaches to the Stars*. Berkeley: University of California Press, 1977.

Steel, Duncan. *Eclipse: The Celestial Phenomenon That Changed the Course of History*. Washington, DC: Joseph Henry Press, 2001.

Steele, John M. *A Brief Introduction to Astronomy in the Middle East*. London: Saqi, 2008.

Whitaker, Ewen A. *Mapping and Naming the Moon: A History of Lunar Cartography and Nomenclature*. New York: Cambridge University Press, 1999.

Williamson, Ray A. *Living the Sky: The Cosmos of the American Indian*. Boston: Houghton Mifflin, 1984.

INDEX

1105887   520 HO $20.00